T0419462

Praise for *Beast in the Machine*

"George Dougherty's eminently readable new work offers deep insights into the implications of robotic combat systems, not only from a technical standpoint with associated challenges and opportunities, but with a moral and ethical view as well. It should be required reading for defense technologists, military historians, or anyone with even a casual interest in the inevitable push toward proliferated autonomous weapons systems."

—Dr. Mark J. Lewis, former US director of defense research and engineering for modernization, and former chief scientist of the US Air Force

"Robotics and AI are having a great effect on battlefields across the globe. Extended ranges and shortened decision times, precision engagements, and increased lethality are but a few of the impacts. Dougherty's review of past applications, assessment of current technology, and projection of future implications provide keen insight into the problems of modern combat.

Beast in the Machine is a must-read for military professionals—those engaged in developing future force capabilities, and those involved in planning and executing current operations. George Dougherty contributes significantly to a most crucial discussion of the future of human conflict affected by the proliferation of robotics and artificial intelligence. America's armed forces cannot afford to be late in adapting to the changes upon us."

—Colonel Don Sando, USA (ret.), former deputy to the commanding general for the US Army Maneuver Center of Excellence

"A rigorously thorough and piercingly insightful look at where today's mil-tech revolution came from . . . and where it's heading. *Beast in the Machine* is a vital book for anyone who wants to understand how the rising tide of robotics and AI will shape twenty-first-century conflict."

—Lieutenant Colonel Dan Ward, USAF (ret.), author of *F.I.R.E.*, *LIFT*, and *PUNK*

"George Dougherty's *Beast in the Machine* provides key insights into the robotic transformation of combat that is now underway. The understanding it provides will be of value to security leaders, defense industry, and warfighters alike. A must-read."

—Lieutenant General David A. Deptula, USAF (ret.), architect of the Desert Storm air campaign and dean of the Mitchell Institute for Aerospace Studies

"This incredible work is a critical current classic that is essential reading for tech warriors, tech bros, and transformative leaders at all echelons because it comprehensively covers the global robotics and AI revolution . . . The nation that leads in the next four years will lead the world in the next two hundred—so read, heed, and lead to WIN!"

—Major General John M. Olson (ret.), former US Space Force chief technology and innovation officer and first Air Force chief data and AI officer

"Even as it's changing our schools and offices, artificial intelligence is transforming battlefields—both literal and political—in new, startling ways. *Beast in the Machine* is a highly accessible guide for making sense of it all."

—Dr. Victoria Coleman, CEO of Acubed, director of DARPA, and chief scientist of the Department of the Air Force

"Drawing from a career at the intersection of technology and warfare, George Dougherty has delivered a compelling set of lessons on the history and future of robotics and its ability to transform warfare. *Beast in the Machine* raises important ethical and policy questions, shares informed practical knowledge, and offers guidance to navigate the future."

—Dr. Arun Seraphin, executive director of the Emerging Technologies Institute for the National Defense Industrial Association

BEAST IN THE MACHINE

How Robotics and AI Will Transform Warfare and the Future of Human Conflict

GEORGE M. DOUGHERTY

BenBella Books, Inc.
Dallas, TX

The views expressed are those of the author and do not reflect the official guidance or position of the United States government, the Department of Defense, the United States Air Force, or the United States Space Force. The discussion of non-federal entities, products, or services does not imply any endorsement by the United States government, the Department of Defense, or the Department of the Air Force.

BenBella Books, Inc.
8080 N. Central Expressway
Suite 1700
Dallas, TX 75206
benbellabooks.com
Send feedback to feedback@benbellabooks.com

BenBella is a federally registered trademark.

Printed in the United States of America
10 9 8 7 6 5 4 3 2 1

Library of Congress Control Number: 2025007186
ISBN 978-1-63774-718-6 (hardcover)
ISBN 978-1-63774-719-3 (electronic)

Editing by Rick Chillot
Copyediting by Scott Calamar
Proofreading by Sarah Vostok and Jenny Bridges
Indexing by WordCo Indexing Services
Text design and composition by PerfecType, Nashville, TN
Cover design by Morgan Carr
Printed by Lake Book Manufacturing

For Vanesa and for Joyce

CONTENTS

INTRODUCTION
Proliferation and Peril

Beginning on September 27, 2020, in the Caucasus Mountains, the world had a glimpse of the future of warfare. At dawn, Armenian soldiers were dug deeply into trenches and fighting positions guarding the disputed territory of Nagorno-Karabakh. They scanned the stark alpine plains and pine-covered mountainsides, waiting.

More than twenty years earlier, after the collapse of the Soviet Union, Armenia had wrested control over the territory by defeating its neighboring country Azerbaijan in a grinding, six-year war. Now tensions were high and Azerbaijan was threatening to take the territory back. The Armenian defenders were confident. The rough terrain favored defense. Plentiful Armenian tanks anchored the defensive lines, dug in behind earthen barriers. The lines were protected by dozens of mobile surface-to-air missile systems equipped with powerful radars. If Azerbaijan's soldiers wanted another fight, the Armenians felt ready. But what followed took them by surprise and would put the world on notice.

The Armenian soldiers manning the air defense radar systems detected enemy warplanes approaching. They fired surface-to-air missiles to intercept them. They were not warplanes, however, but old Soviet biplanes that the Azerbaijani military had converted into drones that simulated warplanes on radar. As the air defense radars remained locked onto the drone biplanes, small Azerbaijani radar-homing loitering munitions, or "kamikaze drones," that were waiting overhead locked onto the radars and dove with a buzzing howl to strike them, turning the air defense vehicles into balls of fire. The Armenian radars struggled to detect these small robotic munitions, which seemed to come out of nowhere. Some of the air defense crews turned their surviving radars off to protect them. Armenian soldiers peered up in dismay as groups of up to thirty enemy drones appeared overhead. Some were delta-winged kamikaze drones, like the eight-foot-long (2.5-meter) Israeli-made Harop (or Harpy). Others included the larger, sleek, airplane-like Bayraktar TB2 attack drones made in Turkey that peered down using high-resolution video cameras, their distant operators identifying the Armenian air defense vehicles by sight and then destroying them with the TB2's compact laser-guided missiles. Devastating blows against the air defenses continued day after day. After two weeks the drones had destroyed some sixty Armenian air defense systems.[1]

As the Armenian air defenses crumbled, the Azerbaijani military turned its robotic weapons on their ground forces. Everything that could be seen from the air was attacked with precision weapons, as battle was reduced almost to a point-and-click exercise. Armed drones and loitering munitions in the air spotted tanks, artillery pieces, and trucks from above and destroyed them where they sat, turning years of investment in traditional military hardware into burning scrap metal. The TB2s dropped laser-guided munitions precisely into the trenches where the Armenian infantry huddled for shelter. The Azerbaijani ground forces used video-guided Spike missiles to add to the precision-guided carnage. Observation drones spotted Armenian units for Azerbaijani artillery and rocket barrages to target and obliterate. The drones captured video footage of it all and fed it live to Azerbaijani control stations, where technicians edited and published it on the internet, further damaging the Armenians' wavering morale.

When the Azerbaijani forces advanced against the shattered defenses, they were led by small units of commandos that remained largely invisible to the Armenian defenders. They advanced stealthily, designating Armenian defensive positions in the forests and hillsides for strikes by long-range artillery and precision-guided missiles. Armenian troops abandoned their positions under fire and fell back, but with each retreat their new positions came under the same deadly precision attacks. The Azerbaijani units advanced until they began to cut off the key cities in Nagorno-Karabakh's central highland valley. By then the Armenian leaders could see the war was lost and quickly negotiated for peace in order to retain what they still could.

The contrast with the previous Armenian-Azerbaijani war could hardly have been starker. After just forty-four days, the bulk of Armenia's armored force was destroyed, with around 150 tanks lost, most without ever firing a shot.[2] Hundreds of other systems were gone as well, and there were thousands of Armenian casualties—soldiers injured or killed. The armed forces were crippled. Despite efforts to save face, the Armenian government was unable to hide the scale of the defeat from its people. Protestors stormed the parliament building and beat the president of the National Assembly. The government nearly collapsed.

This was the outcome of the first war between nations that was dominated by unmanned robotic systems. The shocking victory clearly demonstrated what was possible when robotic weapons technologies now widely available on the international market were combined with a doctrine and concept of operations built to capitalize on their advantages. Importantly, this was not the fantastically expensive act of a world-leading technological power like the United States. It was the work of a tiny country ranked sixtieth in the world for military spending. The conflict proved that with the growing proliferation of robotic weapons, a campaign like that of the Azerbaijanis could be carried out by almost any military power. In its aftermath, militaries around the globe rushed to acquire the new capabilities that the Nagorno-Karabakh war had demonstrated to the world.

One of those countries was Ukraine. Despite being one of the poorest countries in Europe, it bought TB2s from Turkey and accelerated its domestic experimentation with small armed quadcopter drones. In February 2022, not much more than a year after the end of the Nagorno-Karabakh war, Russia launched its full-scale invasion. Ukraine's newly acquired robotic weapons played a significant role in its surprising victory over what had appeared to be an overwhelming initial Russian assault. As the war continued, robotic weapons grew to play a dominant role, confirming to the world that the age of robotic warfare was dawning. By 2024, robotic weapons had become so central to Ukraine's defense that Ukrainian soldiers stated that to win the war, "We don't need three million soldiers, we need three million drones."[3]

This book is an exploration into the past and future of robotics in warfare. Robotics and warfare have a powerful relationship. As history will show, the technical field of robotics was conceived with warfighting in mind. Today, robotic systems and the artificial intelligence, or AI, that increasingly controls them are on the verge of transforming war. The news is filled with reports of strikes by drones and robotic smart weapons—and with worries over the potential consequences of rapidly advancing military AI. Many experts, observers, military personnel, and ordinary citizens alike sense that something big may be happening. That this looming robotics revolution is not just the latest arrival of a new military technology, like jet propulsion or night vision, but a more comprehensive change that could alter fundamental aspects of warfare itself.

This revolution will affect all of us, not only those with an interest in the military or technology. The changes that are coming will impact our lives, our safety, and the kind of world that we and our children will inhabit. Many of our fears about robotics and AI are shaped by cultural influences and inaccurate information. Thus, we tend to worry about the wrong things, and we potentially overlook the actual dangers. Having an informed view will help us all to be part of the conversation around this subject and let us play a role in shaping our future.

Our situation may be akin to that at the start of the twentieth century, when a similar broad-based, comprehensive technological change was starting to sweep the world: industrialization and mechanization. As mechanical engines replaced animal power, marvels such as the "horseless carriage," (the automobile) and the "horseless tiller" (the tractor) appeared. Those first inventions seemed amazing. After all, it was bizarre to see a carriage driving along with no horse attached. But in retrospect, those early mechanized creations were straightforward ideas, simply derived by putting the term "horseless" in front of a concept that was already familiar. However, subsequent military innovations that transformed warfare, such as the tank and the airplane, weren't simply "horseless" versions of things that existed before. They embraced the inherent potential of mechanization to produce revolutionary new systems that could not have existed in pre-mechanical forms.

Today we use the term "unmanned," and alternatives like "unpiloted" or "uncrewed," in a similar way. Many of today's unmanned aircraft, vehicles, and naval vessels are superficially futuristic but are basically unmanned versions of familiar things. What new classes of military innovations may dominate the coming era of robotic warfare that fully embrace the inherent potential of robotics and might never have existed in a pre-robotic form?

Military-technical revolutions aren't just about technology. They also depend on new thinking about how emerging technology might transform warfighting. For instance, the rise of military aviation and the rise of armored warfare were the subjects of revolutionary military thinking in the 1920s and 1930s by innovators like Billy Mitchell in the US, Giulio Douhet in Italy, J. F. C. Fuller and B. H. Liddell-Hart in Great Britain, and Heinz Guderian in Germany. They charted the future course of those developments and how they would change warfare. Those works were informed by the combat experience of early aircraft and tanks in conflicts such as World War I. Militaries that accurately foresaw the implications of the emerging technologies and built their military forces and operational doctrine to capitalize on their advantages—like Germany with armored "blitzkrieg" warfare and the US and Great Britain with strategic airpower—met with success when World War II erupted. Militaries that failed to fully embrace those implications, such as France's, were caught by surprise and suffered defeat.

Innovative thinking is starting to emerge regarding the future implications of robotic warfare. However, this thinking is hindered, among other things, by a poor understanding of its history, even among many of the experts tasked with developing robotic technologies and the military leaders responsible for integrating them into military operations. Many of the most important combat experiences with earlier robotic weapons are poorly documented and nearly forgotten, sometimes because they were initially secret. For instance, the largest tank battle of World War II was spearheaded by unmanned combat ground vehicles. That fact is almost forgotten, as are the lessons that the battle taught about using unmanned vehicles in ground combat. As a result we have been, to a degree, walking backward into the future with our eyes closed. This has already had costly consequences, as small countries and even non-state militant groups have been able to achieve battlefield surprise with innovative uses of robotic technologies that the US and other militaries failed to anticipate.

We are still in the earliest stages of the robotic revolution. Just as robotics and AI are starting to shape civil and commercial life, they are just starting to impact military affairs. All that has occurred so far has only been a prelude to the much more comprehensive changes coming over the next decades. Understanding the important ideas and battlefield experiences that led to the current tipping point can help us to better anticipate and prepare for the even more important changes likely to come soon.

This exploration necessarily involves weapons, because weapons are the tools of warfare. However, simply describing weapons and weapon technologies misses the larger and more important concepts. To paraphrase the physicist Richard Feynman, we can know the names of all the birds but not know anything about birds. Specific weapon systems are like individual data points on a scatterplot. Our goal is to see the entire plot, and to clarify the underlying concepts and trends that determine its shape, so that we can understand why it has the shape it has. Then we can better understand what has happened in the past and predict what may be to come.

The emphasis is on combat. While laboratory experiments and technical demonstrations were important, especially early in the history of military robotics, combat is the proving ground where military ideas and technologies are tested by fire, both figuratively and literally. The most important combat experiences of the past serve as milestones along the road that robotic warfare ideas took to become reality, and they reveal what happened when those ideas were tested in the fire of real military conflict.

Lastly, warfare is more than the use of weapons. It involves other military functions like logistics, intelligence, and command and control. Beyond that, it is ultimately a political activity and a cultural practice. War changes in character over time and from culture to culture, while its fundamental nature as a sociological phenomenon appears rooted in human nature. The robotic transition may shake all those aspects of warfare. This exploration will help present conceptual frameworks that are grounded in an empirical foundation. It will help us to look beyond the present to the day when the inherent characteristics of robotic warfare may drive entirely new ways of warfighting and potentially bring wide-reaching changes to human conflicts and the nature of war.

⊙ ⊙ ⊙

With PhD-level training as an engineer, I worked for many years helping to lead cutting-edge technologies in the Air Force. Partway through that time, I went to a top business school and entered the business world, continuing to serve the military in the reserve. Moving back and forth between corporate boardrooms on one hand, and the Pentagon and military research and development centers on the other, was a unique experience. I was often impressed by the way the chief executives of the high-tech and life sciences companies whom I advised obsessed about the emerging technologies that could disrupt their market positions. Seeming to embody former Intel CEO Andy Grove's warning that "only the paranoid survive," they remembered that their own companies had risen to power and profit by catching a competitor missing a new technology trend, or otherwise overlooking a strategic risk.[4] I and the consulting teams I led worked intensely, sometimes ninety hours in a week, burning the midnight oil in corporate headquarters and crawling to vending machines for late-night sustenance, to help those companies stave off such strategic risks through a new product launch or the corporate acquisition of an innovative startup.

When I changed out of my suit and into my uniform, I couldn't ignore that the same kinds of analyses suggested that the US military might be facing a greater strategic risk, a tidal wave of disruption much bigger than anything my private-sector technology clients faced. A whole array of robotics and AI technologies threatened not only to open opportunities to competitors but to potentially upset the entire "industry," so to speak, of national defense. Yet military leaders were focused on Iraq and Afghanistan, and growing technological threats received little of their attention.

In 2006 I started researching in earnest while off duty, to put together the kind of strategic picture that gave direction to CEOs in analogous positions. I soon found myself spending long hours in archives, speaking with other specialists, studying technical reports, analyzing strategic military documents, and collecting firsthand accounts of combat. The project grew. Technology, history, military science, and business strategy are separate domains, but forging connections between them yielded a story much richer, more colorful, and more complex than I had imagined. It began to clarify the strategic picture just as the most recent conflicts erupted and made audiences, in the Pentagon and in the public, ready for answers.

The term "robot" derives from a Czech word for laborer or serf. The modern usage comes from the 1920 play *R.U.R.* by the Czech playwright Karel Čapek. In the play, R.U.R. is the name of a corporation, "Rossumovi Univerzálni Roboti" or "Rossum's Universal Robots," that manufactures artificial humanlike slaves. They free their masters from the burden of labor. It's worth noting that in the play the "robots" overthrow their human masters and exterminate mankind. Therefore, the trope of robots turning against humans is as old as the term "robot" itself.

In the broadest sense, a robot is an artificial creation that does work that would otherwise need to be done by a human or another living creature. More precisely, robots embody the *core functions* that humans or other living creatures have that enable them to do work. A robot must be capable of three core functions. The first is *sensing*, which is the ability to perceive or take in information about the environment. Sensing can be as simple as detecting a single quantity, for instance temperature, or as complex as high-definition machine vision and other sensory capabilities that mimic or exceed the senses of higher living things. The second essential function is *actuation*, which is the ability to take some external action. This can be as simple as flipping a switch, or as complex as walking on two robotic legs, or operating the controls of a jet aircraft. The third function is *information processing*, which entails interpreting the information from the sensor or sensors and applying some combination of logic or reasoning to decide what kinds of actuation to perform under the conditions. Combine sensing, information processing, and actuation, and you have a robot.

Any machine possessing those three functions can be regarded as robotic, and "robotic" is a quality that exists in degree. An engineered system containing the three essential functions is robotic, but some systems are clearly more fully robotic than others. A smart thermostat meets the minimum qualifications of a robotic system, but it is not as fully robotic as, for instance, an autonomous Mars rover.

Autonomy is the quality of operating without direct human control. Regarding robotics, the term "autonomous" is sometimes used to imply "advanced" and is often described in three stages corresponding to the degree of human oversight required for operation. "Human in the loop" describes the stage where the robotic system cannot act without the command or explicit approval of a human. "Human on the loop" is the stage where the robotic system can act on its own, but only under the oversight of a human who can step in when needed. The final stage, "human out of the loop," describes fully autonomous operation without any human involvement. Autonomy is a good measure of advancement for direct replacement of humans in traditionally human roles, such as piloting vehicles, but isn't in itself a measure of sophistication or success. For example, simple devices like thermostats are already fully autonomous. So is a toddler with a handgun. The toddler is self-directed and exhibits levels of sensing and cognition generations beyond that of most current military robotics. That doesn't mean it would be either militarily effective or a good idea to put a handgun under its control.

Artificial intelligence, or AI, is a term that has been in use since the 1950s and has gone through repeated peaks of interest, excitement, and investment, followed by troughs of disillusionment and disinvestment sometimes called "AI winters." In short, in its first decades it was associated with supposed feats of humanlike reasoning by mainframe computers, which was optimistically assumed to be fairly simple and mechanistic. This was exemplified by early programs like "ELIZA," a text-based virtual therapist developed in the 1960s that could hold simple conversations with a human. To the untrained, it didn't seem far from the imagined capabilities of fictional computer intelligences like HAL 9000 from the 1968 movie *2001: A Space Odyssey*. But it soon became clear that ELIZA and similar programs relied on rudimentary logic and stock phrases and provided no more than a very superficial illusion of humanlike intelligence. It became obvious that any kind of true AI would require profound advances in both hardware and software.

The field came back to life toward the end of the twentieth century by focusing on more well-defined problems and practical applications, such as automatic target recognition and internet search. Now, a recent wave of interest has seen the term AI come fully into professional respectability, as machine learning methods such as deep learning have put whole categories of practical AI applications within the realm of technical feasibility. In the context of robotics, AI is a feature of the information processing function. It enables robotic systems to perform selected perceptual and cognitive tasks that parallel the function of human or animal brains. The level of AI required depends on the degree of authority and autonomy that an application demands, and many military robotics applications such as basic munition guidance require little or no AI. But the advancement of robotics into newer and more challenging applications will most certainly be paced in large part by future advances in AI. Each leap forward in the capability of AI brings new military mission areas within the grasp of robotic systems.

The Nagorno-Karabakh war of 2020, and subsequent conflicts such as the war in Ukraine, are only the most visible examples of looming change. Intelligent machines, capable of sensing their environment, making decisions, and acting on those decisions without direct human intervention, are proliferating into every military mission area. Aerial drones, also known as unmanned air vehicles or remotely piloted aircraft, carry out reconnaissance and strike missions with no human pilots onboard. Unmanned ground vehicles, related to self-driving cars, patrol militarized borders and accompany troops into combat. Robotic underwater vehicles autonomously cruise the oceans for months on end, and unmanned boats surveil harbors for signs of trouble. Ever-more-intelligent smart weapons such as cruise missiles function more and more as fully autonomous robots and carry out a growing share of lethal military attacks.

In the case of the United States military, the growth in the first decades of this century has been striking. In 2001, following the 9/11 terrorist attacks, an American MQ-1 Predator remotely piloted aircraft carried out the first lethal strike by an unmanned aircraft. At the time, only a few Predators existed, for use in airborne surveillance, and the attack was considered novel and controversial. Seventeen years later, the US Air Force was operating twenty-five squadrons of more advanced remotely piloted attack aircraft, and seeking more.[5] The reconnaissance-strike role that those robotic aircraft pioneered has become a mainstay of US air operations. By 2020, approximately 14,000 strike missions had been carried out by remotely piloted aircraft in Afghanistan, Pakistan, Yemen, and Somalia in support of US counterterrorism and counterinsurgency missions.[6]

The US Air Force's remotely piloted aircraft have been among the most visible examples of robotic military power. However, the US Army has amassed thousands of unmanned systems, at first for tactical reconnaissance and explosive ordnance disposal, and increasingly for more complex missions. The expected uses of robotic systems in ground combat are expanding rapidly, with the next generation of even the most mainstream US Army combat systems like infantry fighting vehicles expected to be "optionally manned." Former Chairman of the Joint Chiefs of Staff General Mark Milley predicted in 2024 that within ten to fifteen years, up to a third of the US military will be robotic.[7] As a US deputy secretary of defense put it, "I'm telling you right now, ten years from now if the first person through a breach isn't a fricking robot, shame on us."[8]

Robotic weapons have progressed from being special-purpose tools that complemented more mainstream military systems to being mainstream, frontline systems themselves. During the 1991 Gulf War, known in the military as Operation Desert Storm, US forces used a limited number of "smart" guided weapons to great effect. Those relatively early robotic weapons had an outsized influence on the course of the conflict, delivering pinpoint attacks on the most strongly defended facilities, such as command and control centers, and destroying the highest-value military equipment, such as aircraft protected within hardened shelters. Their dramatic impact was achieved even though they comprised less than 10 percent of the aerial munitions dropped on Iraq during that conflict.

Contrast that to the near-total use of more advanced precision-guided weapons in the highest-profile US and NATO strike missions of more recent years. For instance, on April 13, 2018, the US and its NATO allies conducted strikes that destroyed Syrian chemical weapons sites. The strikes used 105 advanced weapons all fired from outside Syrian territory. All of them were self-guided, terrain-following cruise missiles, including American JASSM-ER, British Storm Shadow, and French SCALP stealth missiles fired from aircraft and a surface warship, plus somewhat older Tomahawk cruise missiles fired from a submarine. None required a human eye on the target. Instead, they navigated to the target areas and avoided air defenses autonomously, finding their final impact points using computerized imaging systems and artificial intelligence.

As another example, the US strike that killed General Qasem Soleimani, the head of the Iranian Islamic Revolutionary Guards Quds Force, outside Baghdad airport on January 3, 2020, was entrusted to an MQ-9 Reaper reconnaissance-strike drone. The use of a remotely piloted aircraft to carry out such a high-profile attack further illustrated how robotic weapons had become a preferred option for important lethal missions.

Robotic weapons have proliferated into the infantry as well. Small, smart weapons like the Javelin anti-tank missile, and NATO analogs like the British NLAW, are giving individual soldiers the ability to destroy enemy armored vehicles at ranges up to several kilometers. They are providing foot soldiers with power previously reserved for tanks and attack helicopters. Hundreds of burned-out Russian vehicles littering Ukraine are testament to their effectiveness.

If robotic systems were only changing the face of operations within the US and its NATO-allied militaries, they would be important. But they are increasingly central to military affairs across the globe. Armed reconnaissance-strike drones similar to the Reaper are no longer unique to the United States. By 2020, manufacturers in China, Russia, Israel, and Turkey were marketing similar large, long-endurance, armed, remotely piloted aircraft. The UK, Israel, Pakistan, Turkey, Saudi Arabia, the UAE, Egypt, and Nigeria all used armed remotely piloted aircraft to conduct lethal strikes between 2015 and 2019.[9] The Turkish-made Bayraktar TB2 drones that achieved fame in Nagorno-Karabakh headlined Ukraine's resistance to invading Russian forces fifteen months later. Lethal, versatile, and inexpensive, armed drones have become a "must have" for global militaries. According to Chris Woods, a journalist who has studied drone use for over a decade, "We are well past the time when the proliferation of armed drones can in any way be controlled . . . So many states and even nonstate actors have access to armed drone capabilities—and they are being used across borders and within borders—that we are now clearly within the second drone age, that is, the age of proliferation."[10]

China, already one of the world's top producers of armed, long-endurance, remotely piloted aircraft, has introduced a new generation of high-performance, stealthy, jet-powered strike drones into service that could surpass fielded systems from the United States and its allies.[11] It is introducing stealthy, missile-armed drone warships.[12] Its industry dominates the global market for small drones and other robots, and the components needed to build them. China is fielding military autonomous underwater vehicles and demonstrating military drone swarms—and has made dominance in artificial intelligence one of its national superprojects.

The country's "Next Generation Artificial Intelligence Development Plan," published in 2017, calls for massive investments to allow China to match the United States in AI capabilities by 2020, and surpass it to become the world's leading AI power by 2030.[13] By 2017, China provided almost half the world's equity investment in AI startups.[14] By 2027, China's investments in AI are projected to exceed $38 billion per year, second only to the US.[15] In addition, China has demonstrated few scruples about applying AI widely to surveillance and security applications. It is working to have a nationwide instant facial-recognition system in place, using thousands of online surveillance cameras and wearable devices, in the next few years. The system's early uses range from locating criminal fugitives to issuing automatic jaywalking citations to citizens. In a darker turn, it is also a key tool in the national social credit score system of behavior control and in the repression of millions of members of the Muslim Uyghur minority in the country's west. China's militaristic neighbor North Korea has also turned toward robotic weapons: in 2023 and 2024 it unveiled domestic reconnaissance-strike drones and copies of the Israeli loitering munitions that had been used in Nagorno-Karabakh.[16]

In 2017, Russian president Vladimir Putin predicted that future wars will be fought by drones, and that "when one party's drones are destroyed by the drones of another, it will have no choice but to surrender." In the same speech, he also described the importance of artificial intelligence, stating his belief that "the [country] who leads in this sphere will be the ruler of the world."[17] The Russian military deployed several models of unmanned fighting vehicles to Syria, tested a next-generation jet-powered stealthy attack drone to work in concert with the Su-57 stealth fighter, and developed an intercontinental unmanned submarine to carry nuclear warheads to enemy coasts. Nonetheless, the Russian military's poor performance against Ukraine's robotic weapons showed that it did not move quickly enough. Its few bright spots of effectiveness during the war were provided by its own growing contingent of precision robotic weapons such as the Lancet loitering munition.

In the Middle East, robotic weapons have proliferated rapidly. Israel was one of the earliest developers and adopters of armed drones and used them extensively, including in its campaigns in Lebanon in 2006 and Gaza since 2008. It has produced innovative small, smart munitions including the loitering munitions and Spike missiles that Azerbaijan used in Nagorno-Karabakh. Israel has also used robotic ground systems such as unmanned bulldozers and robotic dogs in its operations in Gaza.[18] And their investment in robotics has been joined by other regional powers. The Iranian/Houthi precision drone strikes on the Saudi Aramco national oil-processing facility at Abqaiq in 2019 gave evidence that NATO-like strike capabilities are available to second-tier military powers. Both sides of the Libyan civil war operated air forces whose primary strike capabilities were provided by remotely piloted reconnaissance-strike drones. One side's robotic air force was provided by China and the other's by Turkey. This gave them the capability to carry out precision air campaigns at a fraction of the cost of conventional air forces, with the two sides conducting over a thousand precision strikes against each other during 2019.[19]

By 2024 other nations including Ethiopia, Morocco, and Sudan launched their own military campaigns using reconnaissance-strike drones. The proliferation extends to ground robots as well: Pakistan, for instance, has deployed armed ground robots based on Russian designs to its disputed border with India.[20]

The proliferation of robotic weapons is occurring not only in terms of new users but also new roles. US, NATO, and Israeli forces pioneered using remotely piloted aircraft to provide precise close air support to ground forces. By 2019 Israel was using explosive-carrying "kamikaze" drones to conduct destruction of enemy air defense missions over Syria.[21] In early 2020, Turkish forces intervening against the Syrian Army extensively used armed, remotely piloted aircraft in mainstream airpower missions that historically were the domain of manned aircraft. They used robotic aircraft for deep interdiction and attack missions, finding and destroying Syrian mobile air defense systems, armored vehicles, artillery, and infantry.[22] Similar uses were on display in Nagorno-Karabakh and in Ukraine. Until those operations, it was assumed that non-stealthy drones would be of little use in combat against well-equipped armies due to their likely vulnerability to anti-aircraft systems. However, Israel, Turkey, Azerbaijan, and Ukraine have demonstrated not only that such drones can evade air defenses and serve as potent tools against frontline military forces, but that they can be excellent tools for destroying the enemy air defenses themselves. More capable future drones can be expected to expand those roles even further.

Non-state forces have found low-cost robotic weapons increasingly effective and easy to use. Before the Saudi Aramco attack, Yemen's Houthi rebels carried out many smaller attacks of their own against Saudi Arabia, which backed the Yemeni government in the Yemeni civil war.[23] Houthi forces used explosive drones to conduct long-range terror attacks against Saudi airports and other targets. They also deployed Iranian-designed remote-controlled explosive speedboats as anti-ship weapons, scoring a hit against the Saudi frigate *al-Madinah* on January 30, 2017, killing two sailors and injuring three others.[24] Starting in late 2023, Houthi strikes on international shipping using aerial and sea drones and guided missiles have disrupted global trade through the Red Sea, damaging and sinking numerous ships despite extensive efforts at defense by the world's leading navies. Other non-state militant groups like Hezbollah in Lebanon have increasingly operated militarized drones. The Islamic State of Iraq and Syria, or ISIS, innovated the use of off-the-shelf hobby drones equipped with small explosives, such as releasable 40-millimeter grenades, to harass enemy troops.[25] The practice became common during the battle for Mosul in 2016 and 2017 and was quickly adopted by others. Such small armed drones were not "unmanned" versions of anything familiar—they were something new. The difficulty of countering them has prompted an urgent search among leading militaries for better means for defending against small drones, a search that continues today. Having failed to anticipate the innovation, the world's top military forces are struggling to react.

Robotic systems are taking a leading role in modern conflicts around the world, and the distribution, level of capability, and variety of uses of robotic weapons are likely to expand even more dramatically in the coming years. But there are substantial ethical and legal concerns regarding robotic weapons. Numerous scholars have explored the issues raised by the use of unmanned remotely piloted aircraft, such as the US MQ-1 Predator and MQ-9 Reaper, to conduct targeted killings as part of campaigns against violent extremist insurgent and terrorist groups such as al-Qaeda, ISIS, and the Taliban. The concerns that have been raised include the legality of targeted killings as a method in armed conflict; the ethical or political consequences of strikes against different categories of targets and individuals; the potential for noncombatant civilian casualties or "collateral damage" due to mistakes, poor oversight, or technical failures; and the moral hazard of conducting lethal military action from a safe distance. Some of these concerns may decrease with improvements in technology and procedures, but some are likely to persist and grow as robotic weapons become more common.

Other concerned groups and individuals have looked with alarm on the imagined future potentialities of autonomous robotic weapons set loose upon the world and have advocated for an international preemptive ban on lethal autonomous weapons. The United Nations hosted a series of meetings to discuss the prospects for restricting or banning the development of such weapons. The potential danger of technological creations that may act on their own volition is a cultural theme that goes back at least as far as Mary Wollstonecraft Shelley's famous novel *Frankenstein*, published in 1818. More recent incarnations include videos depicting future swarms of miniature airborne "slaughterbot" drones performing mass attacks on the populations of cities.[26] Regardless of the plausibility of any one hypothetical example, anxieties over the potential hazards of autonomous robots and AI have been commonplace in science fiction for decades, to the point of becoming a hackneyed pop culture trope. No one could say that we weren't warned.

The hazards are undoubtedly real. Some particular applications or incarnations may indeed be appropriately subject to restrictions or bans, as was the case for prior generations of weapons. However, the driving forces are real as well, and the readily observable battlefield effectiveness and potential low cost of robotic weapons have produced surging demand. Demographic, economic, and political factors are all motivating the replacement of human warfighters with robots.

Two other specific trends suggest that the continued rise of robotic warfare over the next few decades may be inevitable.

For one, robotics is becoming ubiquitous in everyday commercial and private life, particularly in applications that have obvious kinship to military systems. Hobby drones, autonomous cars and trucks, unmanned aerial and ground vehicles for package delivery, stock-keeping robots in stores, and AI-enabled appliances such as digital assistants that communicate using human language are becoming parts of everyday life. Even systems that don't seem like robots, such as new cars and modern aircraft, are filled with robotic subsystems that do most of the work to monitor, evaluate, and adjust without direct human involvement. It will be commonplace to interact with ever more artificial intelligences and new classes of robots in everyday civilian life. It will be unbelievable if a military strike system doesn't have the artificial intelligence capabilities that are commonplace in consumer systems with which we interact every day. The rising tide of robotics means that expectations for military capabilities associated with advanced robotics and AI will become overpowering.

The second related trend is that essentially all of today's complex systems engineering has embraced the digital robotic paradigm as its default architecture. Sometimes called "embedded systems" or "mechatronics," this means that every class of industrial and consumer product, including transportation systems, utilities, home appliances, and so on, is transitioning from traditional mechanical or analog electrical designs to digitally controlled ones. In these systems, embedded microelectronic processors are fed by sensors that allow each processor to understand the operation of the device and its environment. They're connected to responsive capabilities or actuators that allow the system to adjust or respond appropriately. In short, they are all robotic to an increasing degree.

This robotic paradigm has accordingly become the foundation of training for most twenty-first-century engineers. A robotic or mechatronic system design framework is increasingly the foundation for undergraduate-level instruction in mechanical engineering, aerospace engineering, and other engineering fields. Robotics has even become the predominant framework through which science, technology, engineering, and math subjects are introduced to students in primary and secondary schools, through robotics projects and an ever-growing variety of student robotics competitions. Tasked with the development of future weapons, current and future engineers will assume a robotic framework. Indeed, it will be increasingly difficult to develop a weapon system—or any other complex product—that *isn't* robotic. Attempting to do so may soon be like practicing blacksmithing or glassblowing: the exercise of an antique technology maintained for purposes of nostalgia.

And the pace of technological change is accelerating. Indeed, military robotics is riding the exponential growth curves of the fastest-moving technologies including inexpensive high-performance computing, software for artificial intelligence, miniaturization, electronic sensors, mobile devices, wireless networking, and human-machine interfaces. Those technologies are being driven by colossal global investments for commercial purposes. Some observers have even predicted that the rapid advance of artificial intelligence may lead to a technological "singularity," when AI may overtake human invention and lead to runaway technological advance uncontrollable by humans.[27] Whether that is likely or not, there is little doubt that technology within these key areas will continue to advance rapidly, tending to make robotic weapons and warfare ever more predominant.

As daunting as all that may sound, it's important to remember that while the rise of robotic warfare may carry many risks, it is not a sure ticket to a dystopian future. As we look more deeply into the situation, we will see that robotic warfare emerged from a "scientific" Western military tradition favoring precision, sanitization, and the reduction of unnecessary harm. That tradition has contributed to a significant reduction in worldwide deaths from warfare over the past century. Even the second Nagorno-Karabakh war, for all its lethality, caused fewer than half the military casualties of the earlier, longer, and lower-tech war between the two nations—and only one-twentieth the civilian casualties. As we'll see, the early visionaries of robotic warfare hoped that it might further reduce human suffering from war and perhaps help to bring an end to war overall. There's reason to hope that could ultimately be the case. But it will take wisdom and foresight to get there.

History, unfortunately, shows that high-tech robotic weapons are not necessarily incompatible with barbarism. Real human conflicts are more complex than the push-button warfare imagined by some theorists, and they are likely becoming still more complex in the twenty-first century.

The concepts we'll explore here, with the benefit of history, include how the evolution of future robotic weapons may be shaped by the changing realities and practices of warfare, and how robotic weapons may affect those realities and practices in turn. Rapid global proliferation will complicate those interactions. We will also look at the hazards created by mismatches between human and artificial intelligence, and the false assumptions and human cognitive biases that may amplify those hazards.

In the end, the lesson of all military-technical revolutions has been that military advantages accrue to those who, first, perceive the possibilities inherent in an emerging technology; second, develop effective theory or doctrine that shows how to realize those possibilities; and third, implement and master the theory in practice before their adversaries can. Those steps are also necessary to ensure that the coming robotic revolution, likely one of the most pervasive and important military-technical revolutions in history, and likely impossible to postpone, is navigated wisely.

Wisdom tells us that the US and other leading democratic powers must take a proactive approach to better guide the future development of robotic warfare, anticipate its effects, and manage its risks. That will enable us to escape the reactive mode and the risk of surprise, and to be the masters, rather than the potential victims, of disruptive change. Easily said, but the challenge is huge. The pervasive nature of robotics and artificial intelligence will lead to more implications than can be defined and explored in one place, but we must inspire more thinking along these lines. When cannons and gunpowder arrived in European warfare, they seemed to, in the words of one fifteenth-century observer, "strike as a sudden tempest which turns everything upside down."[28] Ancient kingdoms fell in devastating wars, long-standing political and social orders crumbled, and new powers rose. Our level of understanding, foresight, and preparation may determine whether the changes of the robotic military revolution come as a similar surprise. It may determine if the rise of robotic warfare that is currently underway will favor a more ethical and less violent world, or if it will bring defeat, unintended consequences, and political, ethical, and social upheaval in its wake.

1

RISE OF THE FIGHTING MACHINES

The Secret History of Robotic Warfare

The tank battalions that spearheaded the attack into the Russian lines were equipped with the most modern and high-tech tanks in the army, which recently arrived from the west. However, the Russian forces had spent months building strong lines aof trenches and obstacles, protected by minefields and studded with fortified fighting positions. To help breach those defenses quickly, the tanks were teamed with new, unmanned, robotic assault vehicles. Special "control tanks" were equipped with electronic equipment that allowed each of them to control two of the robotic vehicles. Looking like compact tracked armored vehicles themselves, the unmanned vehicles were fast enough to roam ahead of the lead tanks, forming a skirmish line that sought out and engaged enemy defenses.

The small vehicles weren't just for reconnaissance—they packed a punch. Each carried a 1,000-pound (450-kilogram) bomb that could obliterate the strongest fortifications. Under remote control, they raced up to enemy defenses, zigzagging to dodge enemy fire, and jettisoned their bombs before backing away. The explosions cleared paths through the minefields and tore holes in the Russian defenses, allowing the tanks to advance. Support crews reloaded the unmanned vehicles with fresh bombs, and they raced forward again. When Russian defenders shot at the robots, the control tanks provided covering fire that suppressed the enemy and protected the robots. If a robotic vehicle was knocked out, another simply replaced it. The teaming between tanks and unmanned assault vehicles worked so well that the tank battalions breached line after line of Russian positions, advancing faster than predicted, and soon broke through into the unprotected rear areas behind the Russian forces.

This remarkable operation didn't happen during the Russo-Ukrainian war in the 2020s. Nor was it a scene from a war game or a simulation to test future concepts in manned-unmanned teaming on the battlefield. It happened during the Battle of Kursk, perhaps the largest armored battle in history, on the Eastern Front during World War II, in July 1943.

Robotic combat history remains largely secret, but it has been much more extensive than most of us realize. Sometimes the secrecy was due to actual security classification. For instance, the achievements of the US assault drone units that fought Japanese forces in the Pacific in World War II remained classified until 1966. Even then, most of their records remained uncatalogued in storage until the 1980s, and the units only received belated recognition by the secretary of the Navy in 1990.[29] More often, however, the battles went unrecognized simply because they took place out of sight of the Western press, making no headlines. Sometimes that was the point. Without human pilots or soldiers at risk, militaries could carry out missions quietly. Often, they happened on remote battlefields in places such as the forests of Finland, the steppes of Russia, or the mountains of Afghanistan.

Shining a light on these overlooked clashes does more than illuminate a fascinating aspect of military history. Knowing the history of robotic warfare is important to understanding where we are today, and it teaches classic principles that remain relevant now and in the future. The overall arc of this hidden history reveals two great lessons.

First, military operators have discovered similar advantages and challenges using unmanned systems in combat, regardless of the period. The future innovations that incorporate that combat experience will dominate in the emerging robotic age.

Second, while technology has advanced in ways that the earliest pioneers of robotic warfare a hundred years ago could scarcely imagine, there has been very little advancement in our concepts about the forms that robotic weapons might take and how they should be used. Most of today's widely hyped ideas are incremental improvements on three classical robotic military concepts that the early pioneers established more than a century ago. The failure to advance our concepts along with our technology puts us at risk of surprise when a clever adversary's robotic warfighting innovations suddenly leapfrog our own, something that has already begun to happen.

THE FOUNDING VISIONARIES AND THE CLASSICAL CONCEPTS

In 1898, inventor Nikola Tesla, the father of alternating-current electrical power and many other electrical marvels, launched the technical field of robotics. In response to the outbreak of the Spanish-American War, he built and patented a mobile vehicle that contained electrical elements that he described as analogous to the organs of a living beast. Those elements performed sensing, movement, and control, the basic functions of a modern robot. Tesla demonstrated his invention, which he called a "teleautomaton," at a public exhibition in New York's Madison Square Garden. He wrote that "when first shown . . . it created a sensation such as no other invention of mine has ever produced."[30]

That first robot took the form of a small semi-submersible boat, resembling a miniature ironclad warship. It used a simple, single-channel radio remote control of a kind that might be suitable for a child's toy today. But it was the first of its type anywhere. Audiences were astounded as the teleautomaton moved and obeyed Tesla's commands like a living thing, even using flashing lights to answer simple questions from the audience.

Tesla explained afterward that its intelligence was provided by the remote control, which granted it a kind of "borrowed mind," allowing the "knowledge, reason, judgment, and experience" of the operator to be transmitted to the machine.[31] He stated that "the greatest value of my invention will result from its effect upon warfare and armaments, for by reason of its certain and unlimited destructiveness it will tend to bring about and maintain permanent peace among nations."[32] He predicted that in time, more advanced versions would be created that would increasingly act on their own in response to sensory inputs as if they had their "own mind," able to remember orders given in advance, to record memories, and to make decisions about what to do. In 1900 he wrote of the ultimate impact of such teleautomatic weapons on warfare:

> Certainly, by the use of this principle, an arm for attack as well as for defense may be provided, of a destructiveness all the greater as the principle is applicable to submarine and aerial vessels. There is virtually no restriction as to the amount of explosive it can carry, or as to the distance at which it can strike, and failure is almost impossible. But the force of this new principle does not wholly reside in its destructiveness. Its advent introduces into warfare an element which never existed before—a fighting machine without men as means of attack and defense. The continuous development in this direction must ultimately make war a mere contest of machines without men and without loss of life—a condition which would have been impossible without this new departure, and which, in my opinion, must be reached as a preliminary to permanent peace.[33]

Tesla was an engineer, and his dream of applying robotic technology to war aligned with the modern scientific approach to warfighting that had become increasingly dominant in the West since the Industrial Revolution. However, his proposals for remote-controlled fighting vessels or vehicles that could replace humans in war were rejected by the military authorities of his day. While his ideas were inspiring, his delicate prototypes depended on unreliable handmade components and seemed very far from ready for service on the high seas or in the muck of ground combat.

Nonetheless, Tesla's concept of a remote-controlled fighting vessel inspired many others. One of them was John Hays Hammond, Jr. In 1911, at the age of twenty-three, Hammond built a remote-controlled houseboat using Tesla's teleautomatic principles. He founded an engineering company, the Hammond Research Corporation, perched on the cliffs of the Massachusetts coast, and dedicated it to developing remote-controlled weaponry for naval use. Hammond's main concept was the remote-controlled explosive boat or semi-submersible torpedo for coast defense. While the US Navy didn't adopt the idea in a big way, he received numerous naval contracts, and the Hammond Research Corporation became one of the world's leading sources for radio remote-control components. Hammond achieved other major landmarks in military robotics, including constructing the first full-size remote-controlled battleship, the Coast Battleship No. 4, converted from the former USS *Iowa* in 1921.

Along the way, Hammond and one of his lead engineers, Benjamin Miessner, created an invention that established the second classical robotic weapon concept, following Tesla's remote-controlled teleautomatic vessel. In the early 1900s, warships sailing at night relied on searchlights to navigate and find their targets. Miessner and Hammond reasoned that if an explosive boat could somehow home in on the beam from an enemy warship's searchlight, the boat could guide itself directly to the enemy ship. They tested their solution in 1912 using a small automatic wheeled vehicle they called the "electric dog." It had a pair of selenium photocells mounted on the front like a pair of eyes, with a small divider between them. When light fell on either of the photocell "eyes," circuitry would cause the vehicle's small electric motor to move it forward. If the light fell only on the left eye, the vehicle would steer toward the left, and if light fell only on the right eye, the vehicle would steer toward the right. When the light fell on both eyes, the vehicle would move straight ahead. The electric dog would home in on the beam from a flashlight and follow a flashlight-wielding person around the laboratory like a dog on a leash.

The electric dog established the second robotic concept: the autonomous, self-guiding homing weapon. Miessner saw that he could also apply the same principle to home in on sources of radio waves or sound, in any direction. He saw the possibilities this offered for bringing technological certainty to warfighting, writing in 1916:

> This electric dog which now is but an uncanny scientific curiosity may within the very near future become in truth a real "dog of war," without fear, without heart, without the human element so often susceptible to trickery, but with one purpose; to overtake and slay whatever comes within range of its senses at the will of its master.[34]

The third classical concept emerged from work to improve the safety of early airplanes. American industrialist and inventor Elmer Sperry had invented the electric gyroscope, making gyroscopic autopilots practical for ships. His son, inventor and daredevil pilot Lawrence Burst Sperry, helped create miniature versions that could stabilize an airplane. The younger Sperry dramatically demonstrated his autopilot system to win the 50,000 franc grand prize for safety inventions at a Parisian air show in 1914. Flying a small biplane, he engaged the autopilot and zoomed low and straight past the astonished crowds with his hands over his head and a fearless French mechanic standing on one wing.

The Great War broke out shortly thereafter. Lawrence, influenced by Tesla through mutual acquaintances, became fascinated with the idea of unmanned systems. He realized that his autopilot, in addition to being of great benefit to civil aviation, could provide the key to creating an aerial torpedo, a pilotless, explosive-laden airplane that could fly far into enemy territory to destroy important targets behind the front lines. Marine torpedoes of the day were purely mechanical, short-range devices. Sperry filed a patent for the aerial torpedo in 1916 that included electrical systems for sensing, steering, and control. His electrical autopilot system formed the core. Gyroscopes and barometers sensed the pilotless airplane's altitude and direction, and an engine revolution counter controlled the range. After the engine had turned the correct number of revolutions, the torpedo would dive onto its target. "It is easy to imagine a fleet of these weapons," Sperry wrote, "loaded with deadly gas or explosives, launched against an objective without endangering one human life of the side so employing them."[35] Sperry made a successful test of a prototype in March 1918.

Other inventors around the world also rushed to build autopilot-controlled aerial torpedoes. The most important was a US Army team led by automobile industrialist Charles F. Kettering and an all-star team of automotive engineers, plus Orville Wright of Wright Brothers fame. By applying industrial know-how to the problem, they created a mass-production-ready weapon. It looked very much like a marine torpedo with biplane wings, with the bulging heads of four cylinders in front that drove a propeller in the nose. It resembled an insect, and Kettering called it the "Bug." The team overcame many setbacks, and in October 1918, a successful test prompted an Army report saying, "this new weapon, which has now demonstrated its practicability, marks an epoch in the evolution of artillery for war purposes, of the first magnitude, and comparable, for instance, with the invention of gunpowder in the fourteenth century."[36]

Unfortunately, the aerial torpedo had not truly demonstrated its practicability yet. Both Sperry's torpedo and the Kettering Bug were terribly unreliable and often blew off course due to wind gusts and other disturbances. Sperry realized that a more sophisticated guidance system would be essential if an aerial torpedo was ever to hit a specific target. His team demonstrated a more advanced version that let an observer in an aircraft provide radio course correction. However, with the peace following World War I, military funding dried up, ending both Sperry and Kettering's efforts.

Those three classical concepts from before World War I—the remote-controlled vehicle or teleautomaton, the autonomous homing weapon, and the aerial torpedo—have driven military robotics ever since. The past hundred years can be viewed as a long effort to bring those three concepts to full maturity. The concept of the teleautomaton lives in today's long-endurance strike drones, remotely piloted from thousands of miles away. The homing weapon concept matured into the guided missiles and "smart" bombs that dominate air and naval combat and precision ground attack. The concept of the aerial torpedo has been realized in today's long-range cruise missiles that are launched from air, sea, and land.

Despite their initial limitations, robotic weapons started to affect battle quite early. The first successful combat attack by an unmanned weapon occurred in October 1917, when a German remote-controlled explosive boat hit the British warship HMS *Erebus* off the coast of Belgium.[37] The attack killed two sailors, wounded fifteen others, and left the *Erebus* limping back to port. When World War II broke out just twenty years after the end of World War I, robotic weapons would have major impact.

THE HIDDEN ROBOTIC BATTLES OF WORLD WAR II

Between the world wars, militaries developed remote-controlled aircraft and vessels for target practice. Navies used unmanned ships like Hammond's Coast Battleship No. 4 to practice gunnery, torpedo attack, and aerial bombing. British and US naval engineers built remote-controlled airplanes for anti-aircraft practice. The term "drone" came from that work. The most popular British remote-controlled target aircraft was called the "Queen Bee." Following the British lead, the US Navy developers called their systems "drones." The drone is the mate of the queen bee, and it generally lives only long enough to accompany the queen to a new hive location and fertilize her. As they described it at the time, "to those who know anything about honey bees, the significance of the term will be clear. The drone has one happy flight and then dies."[38]

The British and Americans enjoyed a technological head start, but other countries moved more quickly to embrace the potential of robotics for combat. In particular, the German military had been almost totally dismantled under the Treaty of Versailles that ended World War I, so that when it rearmed under the Nazi regime, it rebuilt with all-new systems and new warfighting concepts. Dreaming of revenge and eager to defeat larger and stronger adversaries, Hitler and the German military hungered for *wunderwaffen*—wonder weapons—that could deliver dramatic results. Its engineers explored unmanned systems in industrial labs and at secret research bases like Peenemünde on the North Sea coast.

The Soviet military also embraced robotic technology at the time, for different reasons. Stalin urgently wanted to modernize the backward Red Army, including equipping it with tanks. He found it easier to manufacture tanks than to produce enough highly trained crews to operate them. To stretch the Red Army's few trained tank crews as far as possible and protect them from battle losses, engineers converted some tanks into unmanned "teletanks." Neighboring control tanks operated the unmanned ones by remote control. When Stalin invaded Finland in December 1939, teletanks were part of the invasion force.

The Soviets found that unmanned tanks could not function like manned ones, especially with the primitive technology of the time. Television wasn't available yet. They armed the teletanks with flamethrowers and machine guns, which needed little accuracy. Still, the control tank crews couldn't aim the remote weapons effectively from their vantage point. Worse, the teletanks struggled in the rugged, snowy terrain of the Finnish forests, getting stuck in ravines or atop anti-tank obstacles hidden in the snow. Most were easily destroyed by Finnish anti-tank gunners. Soviet soldiers resorted to packing the teletanks with explosives and using them as land torpedoes, but most failed to reach their targets.

The Germans, in contrast, didn't try to make unmanned versions of existing vehicles. Instead, their robotic wunderwaffen included remote-controlled precision bombs that let individual German aircraft sink the mightiest warships. On August 27, 1943, a squadron of German bombers used the new Hs 293 winged, radio-guided, rocket-propelled bombs to destroy the British warship HMS *Egret* and damage the Canadian destroyer *Athabaskan* off the coast of France. The Allied ships never got to fire a shot before the weapons struck home. Two weeks later, the bombers attacked the Italian battle fleet as it attempted to surrender to the Allies. With heavy armor-piercing Fritz X bombs that used the same radio guidance as the Hs 293, they sank the 45,000-ton Italian flagship *Roma* and damaged her sister battleship *Italia* while remaining outside of anti-aircraft range.

Next, the Luftwaffe turned the bombers against the Allied invasion fleet off Salerno, where they did so much damage that the invasion of Italy was almost stymied. Soon after, a hit from an Hs 293 sank the British troop ship *Rohna* in the Mediterranean, loaded with thousands of US soldiers, killing 1,149 soldiers and crew and resulting in what remains the greatest-ever loss of American life at sea due to enemy action, a disaster that remained secret until the 1960s.[39] The Allies' saving grace was that the Germans only had a few of these weapons, and only two specialized bomber units could use them.

The Germans also took the aerial torpedo concept to war on a massive scale. When the Allies established air superiority and British and American bombers began pounding German cities and industry, Hitler demanded a means of striking back. The Luftwaffe chief of development and production Erhard Milch and engineer Robert Lusser hatched a secret program to obliterate London using jet-propelled aerial torpedoes. Hitler named his first *Vergeltungswaffe,* or vengeance weapon program, the V-1. They built a massive network of launch sites and mass-production facilities to launch hundreds of V-1s per hour and reduce London to burning rubble in a matter of days.[40] It could have then blasted city after city off the map. Only a massive preemptive campaign of aerial bombing against the V-1 production and transportation network, kept secret to prevent panic by British civilians, averted catastrophe. The campaign accounted for 15 percent of all US bombs dropped on Europe during the war, and the US lost 450 planes in the effort.[41] The strange weapons that began falling on Britain in June 1944 still caught worldwide attention. Variously called buzz bombs, robot bombs, robombs, or doodlebugs, the Nazis launched ten thousand of them against England over four months, and more than 2,500 struck the London area. While harrowing, it was only a few percent of the massive onslaught the Germans had planned. Luckily, the Germans hadn't perfected effective guidance for the V-1s, so the two or three per hour that got through fell randomly over the London area as terror weapons.

The Germans also fielded teleautomatic wunderwaffen for ground combat, and they took an innovative approach. They selected the quirky German automobile manufacturer Borgward, known for its unique luxury sedans and handy three-wheeled Goliath utility trucks, to design a remote-controlled vehicle for clearing mines. The company dubbed this early model the B I. Soon Borgward became more ambitious and developed advanced models capable of different missions. They developed the Goliath tracked mine, a wire-guided mini tank containing 100 pounds (45 kilograms) of explosive, for the German pioneer engineering troops to use in clearing minefields and enemy obstacles. Then they unveiled Goliath's high-end counterpart, the Borgward B IV, for use by German panzer units. The B IV wasn't intended to replace tanks but to complement them, providing them a powerful demolition capability to defeat heavy fortifications at long range. A detachable trapezoidal container at the front of the vehicle contained the heavy explosive charge. German units used B IVs to help breach the defensive fortifications around the Soviet Black Sea port of Sevastopol. Afterward, they published guidelines for employing the B IVs that encouraged their additional use as reconnaissance vehicles roaming ahead of armored spearheads to uncover minefields and reveal enemy positions by drawing defensive fire.[42] The B IV was what we now call "optionally manned." In combat it was remote-controlled, but between battles, it could be driven by a soldier like a normal armored vehicle, making it convenient to bring on campaigns. The vehicle could travel at the same speeds and over the same terrain as the tanks. Its remote controls were simple and easy to use, just a box with a joystick and several buttons.

At the Battle of Kursk, described at the beginning of this chapter, three panzer companies equipped with control tanks and B IVs accompanied the mighty Tiger tanks and Ferdinand tank destroyers that led the German attack. Their ability to be sacrificed when convenient, a quality known today as "attritability," let them conduct attacks no manned vehicle could undertake. They performed well on the first days of the battle, especially when teamed with the Tigers. If the supporting panzer divisions had been in position to exploit their breakthrough, they might have become famous for helping to win the greatest tank battle in history.[43] As it happened, the Soviets had time to close the gap in their lines, and there weren't enough B IVs to sustain operations longer than a few days. After a titanic struggle, the Soviets won the Battle of Kursk, and the achievements of the robotic vehicles were buried amid the greater story of the German defeat.

History wasn't done with the B IV yet, however. They fought in many smaller battles as the German military withdrew under pressure of the Allied offensives. A critical chapter of their story burned them, and the Goliath, into the memories of the Polish people. When the Soviet Army approached Warsaw in August 1944, the Polish citizens launched a revolt and fought the German forces occupying the city. Rather than moving in to help the rebels, Stalin's army waited while an enraged Hitler gathered every unit he could find to crush the Warsaw Uprising. The Polish Home Army had thousands of volunteer fighters in strong defensive positions, and they were fighting for their freedom. Hitler decided to use extreme brutality to make up for numbers. He threw in his wunderwaffen, including B IV panzer units and scores of Goliaths. He also committed brigades of his most brutal SS troops, who had spent the previous years burning Soviet villages and filling mass graves with murdered civilians in occupied territory. Even other German units were appalled by their savagery. German soldiers used B IVs and Goliaths to blast holes in Polish street barricades and buildings, and the SS troops swarmed in to engage the Polish defenders in close-quarters street fighting. Where they took ground from the Home Army, they engaged in rampages of looting, rape, torture, and mass murder. Between 100,000 and 200,000 civilians died while the Nazis slowly crushed the uprising. The presence of the high-tech robotic weapons did nothing to sanitize the brutality. A monument on Kilińskiego Street still bears a piece of track from a B IV, memorializing the victims of one explosion that killed three hundred defenders and wounded hundreds more.

The German wunderwaffen were the face of robotic weaponry through most of World War II. However, after the Japanese Navy bombed Pearl Harbor, the US entered the war and turned its full technical and industrial might toward weapons development. Secret programs gradually restored US leadership in robotic technology. In early 1942, the US Navy was reeling from the Japanese advances in the Pacific and seeking all available means to counter the onslaught. Lieutenant Commander Delmar Fahrney, the leader of the Navy's aerial drone programs (and co-inventor of the term "drone"), and his engineers had developed target drones that could perform realistic attack maneuvers like dive-bombing. This led Farhney to consider the possibility that the Navy could use drones to conduct actual attacks on enemy ships. His team quickly performed some tests using dummy weapons, which confirmed that the idea worked.

The chief of naval operations allowed Fahrney and Captain Oscar Smith, an aviator in the chief of naval operations office, to design a secret program to develop drones for combat. Their initial plan called for operating high-performance assault drones from aircraft carriers and other warships, potentially including merchant ships converted into drone carriers. It called for eighteen squadrons of assault drones and drone-control planes.[44] The Navy brass realized this was too ambitious for an unproven concept but approved a more limited program that created Special Tactical Air Group One, or STAG-1, which took radio-controlled drones into combat.

Because all American aircraft carriers were already in high demand, STAG-1 had to operate from land and also had to develop purpose-built drones that didn't compete with conventional warplane production. In response, Fahrney and his industry contractors developed the TDR-1 assault drone, a sleek, inexpensive twin-engine airplane that could carry up to 2,000 pounds (900 kilograms) of weapons. It could take off and land like a conventional plane. Like the B IV, it was optionally manned, so a pilot could fly it normally between bases. When preparing for an attack mission, ground crews replaced the clear cockpit with a flat cover and activated the radio controls. In addition to an autopilot and radio-control equipment, it contained an early black-and-white TV camera in its nose that transmitted live images. Four-man crews in specially-modified Avenger torpedo bombers controlled the drones from miles away.

After stateside training, STAG-1's personnel and equipment sailed for the Western Pacific aboard the escort carrier USS *Marcus Island* and two Liberty ship transports on May 18, 1944.[45] They disembarked at Banika Island airfield near Guadalcanal, where they established flying operations. On July 30, they demonstrated their readiness by conducting live weapon attacks against a beached Japanese freighter left over from the naval battles there in 1942. Peering at their TV screens, the controllers in the Avenger control planes guided four assault drones, each carrying a live 2,000-pound bomb. They performed first-person video, kamikaze-style attacks, resulting in two direct hits and two near misses.

By that point in the war, conventional US naval power was overwhelming the Japanese, and the Navy leaders no longer felt they needed the assault drones. The Army had just canceled a similar project in Europe called "Operation Aphrodite" that tried to use drone-converted bombers packed with explosives to attack heavily fortified targets, with little success. Nonetheless, they consented for STAG-1 to expend its existing drones and weapons in combat against the giant Japanese naval base at Rabaul. The Japanese had evacuated their ships and aircraft, but thousands of Japanese troops still held the port and the surrounding military installations. STAG-1 flew to a new base 450 miles to the northwest of Banika Island, within range of Rabaul, and conducted a month of strike missions from late September until late October 1944.[46] They set milestones for robotic combat that wouldn't be matched for decades.

STAG-1 began by attacking Rabaul's anti-aircraft defenses. Using TV-guided kamikaze-style strikes, they destroyed a grounded merchant ship that the Japanese had converted into a platform for anti-aircraft guns and knocked out several anti-aircraft batteries on land. It was the first use of drones in the suppression of enemy air defenses role. Then they struck supply dumps, bridges, and other targets including a lighthouse. When the drone operators couldn't get a clear TV image of their intended target, they searched for and attacked secondary targets. By late October they moved beyond simple kamikaze-style attacks and were attacking two targets per sortie, dropping half of a drone's bomb load on one target via radio command and then using the remainder in a kamikaze attack on a second target. Despite the slow speed and vulnerability of the TDR-1 aircraft, the success rate for drone attacks approached 50 percent.[47] That was more than ten times better than the most precise conventional naval attack aircraft, such as dive bombers.[48]

Fahrney and Smith reported that they were able to duplicate the effectiveness of Japan's kamikaze suicide attacks without endangering any pilots. US Navy and Marine commanders in the Western Pacific certified that STAG-1 performed admirably. Nonetheless, the electronics were still unreliable, and the drones required more time and preparation than conventional aircraft. The primitive TV had poor resolution and contrast, meaning that the operators could only see large objects silhouetted against a much darker or lighter background. The top Navy leaders decided they would prefer precision robotic bombs that entailed less burden and could make its manned carrier planes more effective, like those the Germans had fielded.[49] At the end of October the Navy canceled the project. To maintain secrecy, the Navy ordered STAG-1 to bulldoze its control planes and surviving drones into the sea.

The focus on precision bombs yielded fruit. In the final months of the war the Navy fielded the "Bat," a winged radar-homing glide bomb for attacking ships from heavy patrol bombers. Despite the limitations of its primitive analog electronics, it scored several hits on enemy ships before the end of the war. The US Army developed its own radio-guided bombs. It also reverse-engineered the V-1, producing an improved version with rudimentary radio guidance called the JB-2. American companies manufactured over a thousand, and many were on their way to take part in the assault on Japan when the atomic bombs ended the war.[50]

The experience of World War II solidified the three classical concepts. The radio-controlled teleautomaton had proven itself in combat, both on land in the form of the B IV and in the air in the form of the assault drone. Both had shown the value of qualities like attritability and optional manning. Technology was just starting to make homing weapons like the Bat practical. And the aerial torpedo had entered combat on a mass scale, in the form of the V-1. It was clearly going to be a major weapon of war, if some form of intelligence could guide it more precisely to a target.

As it happened, secret wartime projects breaking enemy codes and designing atomic bombs produced a new technology that would help do just that: the electronic digital computer.

FROM THE COLD WAR TO DESERT STORM

When the Cold War began, computers were large, fantastically expensive, unreliable machines based on vacuum tubes. When it ended, the desktop PC was based on microchips and was thousands of times cheaper and more powerful. Computer-controlled robotic weapons followed the same trajectory, evolving from exotic and expensive special-purpose systems into mainstream tools that dominated combat by the superpowers and their client states. As the enabling electronics and software technology improved, the weapons became more reliable, smaller, more capable, easier to operate, and more affordable. The process may seem straightforward and predictable in retrospect, but in practice every time improving technology allowed one of the three classic concepts to be applied effectively within a new domain of combat, it shocked the military world.

At the start of the Cold War, it appeared that the advent of the atomic bomb had made conventional war obsolete, and any clash between the two new superpowers would focus on mutual bombardment with nuclear warheads at long range. The jet-powered pilotless aircraft or aerial torpedo that the Germans had just demonstrated in the V-1 campaign seemed the perfect means for push-button, atomic-age warfare. As the US military's influential 1946 Von Kármán report put it:

> A part, if not all, of the functions of the manned strategic bomber in destroying the key industries, the communication and transportation systems, and military installations at ranges of from 1000 to 10,000 miles will be taken over by the pilotless aircraft of extreme velocity.[51]

The US rushed to field aerial torpedoes, now called "pilotless aircraft," for nuclear missions. The newly created US Air Force established two pilotless bomber squadrons in Europe in the early 1950s, equipped with the new B-61 Matador pilotless nuclear bomber.[52] It was a large, unmanned jet aircraft carrying a nuclear warhead. It launched from a short ramp using strap-on rocket boosters that accelerated it until its jet engine took over. By triangulating the signals from friendly radio beacons on the ground, it could navigate to within a few miles of its target over a range of several hundred miles, close enough for atomic weapons. The US Navy fielded a similar pilotless nuclear bomber, the Regulus, that it launched from submarines. Subsequent 1950s programs worked to develop pilotless aircraft that were bigger and faster, with intercontinental range. Other programs rushed to build pilotless aircraft that would defend the US homeland from Soviet nuclear bombers, whether piloted or robotic. The Ground-to-Air Pilotless Aircraft, or GAPA, project followed late-war German work on aerial torpedoes that could ram enemy aircraft, and that led to giant surface-to-air missiles such as the Nike and Bomarc series. In the E-ring of the Pentagon, where the most senior defense leaders and their staffs work, there remains a 1952 painting entitled *Air Defense*, depicting futuristic robotic jet aircraft ramming early Cold War–era bombers.

The Soviet Union embraced pilotless aircraft with equal fervor, though emphasizing different goals. It hurried to test its own atomic bomb while building a massive network of surface-to-air missiles to counter the US bomber advantage. As a result, the Soviet military enjoyed a lead in surface-to-air missiles throughout the Cold War. Lacking the powerful aircraft carriers of the US Navy, the Soviet Navy also went all-in on pilotless aircraft as its means of countering Western naval superiority. Soon, Soviet naval ships bulged with the launch hangars for pilotless aircraft, both nuclear and conventionally armed. The Soviet Navy and Air Force fielded missile-launching bombers with still more varieties of pilotless aircraft slung beneath their wings, often adapted from early Soviet jet fighters.

The problem with all of this for both sides was that the clunky vacuum-tube electronics technology of the 1950s was unable to meet the futuristic expectations. While they looked impressive, the reliability of those big, unmanned weapons was terrible. For instance, the first US intercontinental pilotless bomber, the Snark, had a guidance system that weighed 1,500 pounds (650 kilograms) and relied on a second, huge computer at the launch site for its guidance calculations.[53] Launch sites were spectacularly expensive and required crews laboring around the clock to keep them in working order. Despite that burden, in tests only one in every three Snarks even made it off the launch ramp, and only one in ten reached its target.[54]

Meanwhile, conflicts such as the Korean War proved that conventional, non-nuclear war wasn't obsolete after all. Manned bombers continued to be useful and more reliable than unmanned aircraft. Ballistic missiles, which were simpler and had less complex guidance needs, began to arrive and supplanted winged ones in the nuclear role. As a result, by the late 1950s the early mania for "pilotless aircraft" ended.

However, robotic weapons overall were taking off. They migrated from the strategic nuclear regime into the tactical regime. William McLean and his team at the US Navy's China Lake development center pointed the way. In 1955 they tested their Sidewinder heat-seeking missile, a true self-guiding homing weapon just as envisioned by Hammond and Miessner forty years earlier. It homed in on the bright infrared spot created by a jet's exhaust. Importantly, McLean's team decided that robotic technology should make a weapon simpler to use, not more complicated. They built the Sidewinder using the body of a simple unguided rocket. All a pilot had to do to employ it was check that the missile "saw" the bright heat spot from the enemy aircraft and launch the missile. The missile did the rest. It greatly outperformed the other early missiles of its day. When US engineers equipped Taiwanese Sabre jet fighters with the new missiles during the Taiwan Strait crisis of 1958, the Sabres devastated gun-armed Chinese MiG-15s and ushered in the age of aerial missile combat.

Guided missiles thus became smaller and less airplane-like, and militaries started to think of them more as smart munitions than as pilotless aircraft. As solid-state transistors replaced vacuum tubes, they became more reliable and less burdensome to maintain. During the 1950s and 1960s, homing weapons scored more and more dramatic successes. One year after the Sidewinder's debut, a Soviet-made SA-2 scored the first victory by a surface-to-air missile, when Chinese forces shot down a Taiwanese reconnaissance jet. Missiles so quickly dominated air combat that by the 1960s new fighters such as the American F-4 Phantom omitted guns entirely. Then in 1967, small Egyptian missile boats armed with Soviet radar-homing SS-N-2 Styx anti-ship missiles surprised and sank the Israeli destroyer *Eilat*, demonstrating that small vessels armed with missiles could destroy even large warships. Missiles quickly replaced guns on naval ships, and first-generation airplane-like missiles like the Styx gradually gave way to smaller ones that no longer resembled aircraft.

The 1973 Yom Kippur War between Israel, Egypt, and Syria was a watershed moment. Egyptian surface-to-air missiles shot down nearly a hundred of Israel's American-made aircraft and made Egyptian territory almost immune to air attack. Militaries around the world considered that the end of the manned warplane might be at hand. Even more shocking, Egyptian infantry decimated Israel's vaunted tank forces using small anti-tank guided missiles. Maxims like "the best enemy of the tank is another tank" seemed suddenly obsolete. With tremendous effort and emergency supplies of missiles from the US, Israel was able to turn the tide, but its near defeat left it shaken. The only domain where it had enjoyed superiority was at sea, because after the *Eilat* incident it had replaced all its navy ships with missile boats.

The dominance of homing missiles had one big limitation: In many military situations there was nothing useful for a missile to home in on. Heat-seeking missiles homed in on a jet's hot exhaust. Radar-homing missiles homed in on the radar reflection from an enemy plane or ship, lit up by friendly targeting radar the way a flashlight illuminates a flying moth at night. Special anti-radiation missiles could also home in on the signal emitted by an enemy's radar. But for most other targets, especially on land, there was no high-contrast signal. American engineers overcame that limitation with the help of a new technology, the laser. In 1965, US Air Force and Texas Instruments engineers demonstrated that they could use a laser to put a bright spot on anything at all—and then use a laser-guided bomb to home in on that spot. The homing mechanism was just an updated version of the one demonstrated by Hammond and Miessner. By 1972, US planes in Vietnam were using laser-guided bombs to destroy bridges that had survived dozens of airstrikes using unguided bombs.

However, when the US Air Force sent its new laser-guided wonder weapons to raid the complex of jungle roads known as the Ho Chi Minh Trail, it found that its new problem became finding targets to attack. Trucks and supply dumps amid the trees were much harder than bridges to find from the air. The Air Force realized that precision weapons would require lots of precision targeting information.[55] As a result, the US military launched many programs to find new technologies to locate targets for precision attack. Thus arose the modern field of tactical intelligence, surveillance, and reconnaissance, abbreviated as ISR.

The rising lethality of homing missiles caused a resurgence of interest in pilotless aircraft. The US lost several manned reconnaissance planes to Soviet SA-2 missiles in the early 1960s, sparking major international incidents. In response, the US Air Force secretly developed reconnaissance versions of the BQM-34 Firebee jet-powered target drone. The small, fast, and elusive Firebees proved able to sneak past anti-aircraft defenses and get in close to photograph sites wherever needed, without putting any pilots at risk. Launched from the ground like the Matador or carried under the wings of DC-130 transport aircraft that also served as their drone control planes, secret Air Force units sent them on missions over communist China and then over Vietnam. Despite using film cameras, they were able to get unprecedented amounts of up-to-date tactical ISR information. As their handiness became clear, the Air Force began to use them for an increasing variety of special tactical missions. By the early 1970s, the Air Force was equipping them to conduct electronic warfare and had demonstrated their ability to launch air-to-surface homing missiles to destroy ground targets.[56] The Navy launched them from ships, including from an unmanned surface vessel.[57] The Navy's "Top Gun" weapons school even used them briefly as aggressor aircraft, where they outmaneuvered veteran instructors in F-4 fighter planes.[58]

After Israel's shocking losses of American-made warplanes to surface-to-air missiles in the Yom Kippur War, American think tanks proposed in the 1970s that high-performance jet drones were the future of airpower. Able to get in close, they could hit air defenses and other targets accurately, and their low cost and attritability made losses acceptable. To make them truly effective, they would have to be "remotely manned" by pilots who would have the sensation of being inside the drones via real-time, high-definition video.[59] The RAND Corporation recommended a force of reusable drone combat aircraft, each costing about a tenth as much as leading manned fighters, plus a larger number of one-way "kamikaze" drones ten times cheaper still.[60] Unfortunately, the television and data networking technology of the time was still far too primitive. Immature technology, and the deep budget cuts that followed the end of the Vietnam War, once again postponed the rise of high-performance teleautomatons.

Israel, however, did not give up. Despite its smaller budgets, it developed inexpensive propeller-driven drones that carried high-resolution video cameras. It used them during the 1982 Lebanon War. By using the attritable drones in partnership with strike jets, the Israeli Air Force was able to destroy all nineteen Syrian surface-to-air missile systems in the strategic Bekaa Valley without the loss of any manned jets. Though lower tech than the American think tanks had envisioned, remote-controlled drones had once again proved their combat effectiveness, and Israel became a world leader in military drones. Engineers kept drone work alive in the US by researching similar low-tech "harassment" drones to jam or attack enemy air defenses.

Meanwhile, anti-ship homing missiles became more and more lethal. Some missiles carried their own targeting radars, so they could guide themselves to their targets autonomously and even find their own targets. During the Falkland Islands conflict in 1982, Argentina had only six French-made Exocet homing missiles but scored four hits against British ships, sinking two and damaging a third.

Advancing computer technology finally allowed aerial torpedoes to achieve the vision of their original inventors. In the 1980s, the US fielded cruise missiles such as the Tomahawk. Truly resembling a flying torpedo, the jet-powered Tomahawk could fly up to a thousand miles to precisely strike designated targets. Engineers addressed the missing piece of the aerial torpedo puzzle, precision navigation, by using onboard computers to compare optical or radar images of the terrain below the missile with prerecorded overhead images along the intended course. The missile could fly from waypoint to waypoint, using the recorded images to correct its path as it went. As it approached the target, the images became closer together and higher in resolution until the missile merged with the target and detonated a 1,000-pound (450-kilogram) warhead. Later, satellite-based global positioning system, or GPS, navigation let cruise missiles navigate precisely, even over areas with featureless terrain like deserts and oceans.

Lastly, late in the Cold War, the first autonomous loitering munitions went into development, such as the Tacit Rainbow, a jet-powered winged missile that could patrol over enemy territory for lengthy periods, scanning for air defense radars and then diving to destroy them. Other concepts followed later, such as the Low-Cost Autonomous Attack System, or LOCAAS, for finding and attacking ground targets. However, immature technology, high cost, and the end of Cold War funding halted those programs.

Late Cold War NATO strategies envisioned using masses of precision weapons to defeat a Warsaw Pact invasion of Western Europe. Fortunately, the peaceful collapse of the Soviet Union meant that all those Cold War weapons were never used in a superpower conflict. However, before major arms reductions occurred, Saddam Hussein of Iraq chose to invade Kuwait. All the new weapons the West had fielded to defend against a Soviet attack on Europe were instead used to expel the Iraqi military from Kuwait during Operation Desert Storm.

The robotic weapons awed the global TV audience. News correspondents reported from hotel balconies in Baghdad while Tomahawk cruise missiles roared past in the night and laser-guided bomb strikes lit up the sky. Infrared video footage showed bombs precisely striking targets at will, from hardened aircraft shelters to the Iraqi air defense headquarters. The so-called "smart bomb" became the iconic weapon of the war. Behind the scenes, drones and aerial decoys spoofed Iraqi air defenses. Unseen to most observers, powerful new ISR systems like surveillance satellites and the Joint STARS airborne ground-surveillance radar planes, backed by small armies of intelligence analysts, produced targeting data for those smart weapons. Using that data, pilots devastated Iraqi armored vehicles. One Iraqi soldier who had fought in his country's earlier war with Iran stated that his armored unit took heavier losses in thirty minutes of allied air attack than during the eight years of the Iran-Iraq War.[61] Coalition tanks using robotic fire-control systems blasted the surviving Iraqi tanks like ducks in a shooting gallery. Only 9 percent of the aerial weapons were "smart" robotic munitions—and fewer still of the ground weapons. But when the war was over, the Iraqi military was smashed, and out of the almost 700,000 US personnel that were deployed, fewer than 150 were killed, a rate not much higher than might be expected from accidents during a live-fire training exercise.

Desert Storm showed how much traditional battle had changed, at least for the richest countries. The combination of powerful ISR with computer-controlled smart weapons enabled so-called "parallel operations," where vast numbers of targets could be hit precisely at the same time, causing shock similar to that of a nuclear attack but with vastly greater discrimination. On the first day of Desert Storm, coalition aircraft attacked 152 targets, more than the total number of targets hit by the strategic bombers of the Eighth Air Force in Europe during all of 1942 and 1943.[62] With such precise weaponry and enough information, an enemy military or nation could be taken down in an engineered fashion, much like a demolition company brings down a tall building in a controlled implosion. Worried leaders in China, Russia, Iran, and elsewhere resolved to get similar capabilities as quickly as possible.

TECHNOLOGY AND TERROR

After the terrorist attacks of September 11, 2001, the US and its allies shifted their attention to twenty years of bitter combat against terrorist and insurgent groups, starting in Afghanistan, then in Iraq, and then in the territories seized by ISIS. Counterterrorist and counterinsurgency operations extended to Yemen, Somalia, the Sahel region of Africa, and elsewhere. In those unconventional conflicts, the enemy had few important facilities or even armored vehicles. Large smart bombs and cruise missiles were not much use. If Desert Storm had suggested that push-button war had become possible, the experience of the "War on Terrorism" put that notion to rest. Instead, combat operations featured special-forces raids, operations in crowded cities, and intensive surveillance, often requiring the ability to track and observe even individual persons and then strike very precisely when the right moment or target arrived.

The slow, propeller-driven reconnaissance drones with high-definition video capability that Israel had fielded proved ideal for those missions. The US had operated a few Israeli-designed Pioneer drones during Desert Storm. During the UN intervention into the Balkan Wars in the 1990s, the US used a handful of Predator drones, created by the Israeli designer Abe Karem, to provide live high-definition video of conflict areas. The service rapidly became indispensable.[63]

In 2001, the US Air Force and the CIA upgraded Predators with satellite datalinks and Hellfire anti-tank missiles and sent them to Afghanistan to hunt for Osama bin Laden and other al-Qaeda and Taliban leaders. An unexpected incident showcased their value in other combat operations. When US Army Rangers and Navy SEALs landed in the Afghan mountains to fight al-Qaeda on March 4, 2002, they became stuck in a fierce battle atop a high ridge. The militants shot down one of their helicopters, and soldiers on the ground were pinned down by fire from an al-Qaeda machine-gun bunker hidden under a rock outcropping. Several air strikes failed to hit the bunker. A Predator happened to be flying nearby. Its crew, who were actually sitting thousands of miles away in a ground-control station in Northern Virginia, offered to help. The Predator crew observed the bunker using high-definition video, then placed a laser spot right in the opening of the bunker and sent a laser-guided Hellfire missile into it. That turned the tide. The Predator stayed on station throughout the remainder of the Battle of Roberts Ridge, providing real-time surveillance and lasing targets for laser-guided bombs dropped by many other aircraft.[64] Henceforth, armed overwatch by remotely operated reconnaissance-strike drones became a standard part of coalition ground operations.

The US military began the War on Terrorism with fewer than fifty unmanned aircraft in its operational inventory. By early 2010 it had thousands. Most were small ISR drones, but the long-endurance, armed drones like the MQ-1 Predator became the iconic weapons of the War on Terrorism. With them, the remotely manned aerial teleautomaton became a mainstream weapon. Building on the assault drones of World War II and the Firebees of the Vietnam era, they added modern control technologies based on high-definition video and high-bandwidth satellite networking that allowed them to be remotely piloted from anywhere on Earth. The Predator was joined by more advanced models such as the MQ-9 Reaper, refined for combat.

Waves of new tech from the booming mobile device industry helped shrink robotic systems. They included gyroscopes integrated onto microchips, miniature GPS receivers, and tiny high-definition video cameras. Even small hobby drones acquired the memory and computing power to process real-time digital video and run sophisticated automatic flight controls. After 2010, machine learning techniques using artificial neural networks started to enter widespread application, enabling leaps in artificial intelligence.

All this cheap capability made it possible to combine some of the features of the classic robotic weapon concepts. New loitering munitions like the Israeli Harop combined aspects of remote-controlled drones, homing weapons, and long-range aerial torpedoes. By the Russo-Ukrainian war, even compact first-person-view, or FPV, racing drones could be armed, yielding loitering munitions of handheld size.

The focus on ground-based counterinsurgency warfare inspired a belated revival of interest in robotic ground systems. The US Marines had fielded some remote-controlled tanks to assist with breaching Iraqi sand berms in Desert Storm, but even that modest echo of the robotic ground operations of World War II had been canceled at the last moment.[65] Now the Army and Marines fielded hundreds of small remotely operated ground robots for defusing unexploded bombs and providing video surveillance within buildings and confined spaces. Most were small tracked vehicles that resembled the German Goliaths of World War II, but with the explosives replaced by video cameras and robotic arms.

The terrorists and insurgents fielded their own sorts of poor-man's robotic weapons that let them challenge the much more advanced coalition forces. The improvised explosive device, or IED, often detonated by remote control, became the insurgents' deadliest weapon, producing around half of all US casualties. After ISIS swept to power in 2014, its improvised drone air force vexed Iraqi and supporting US forces during the battles for Mosul and Raqqa and quickly inspired similar tactics around the world. Examples of sophisticated strikes by foreign kamikaze drones and loitering munitions, such as the Iranian and Houthi precision attacks on the Abqaiq and Khurais oil facilities in Saudi Arabia in 2019, showed that the rest of the world was using the new tech to craft new weapons.

Western militaries moved too slowly, either to make the classic robotic weapon concepts more affordable and plentiful or to move beyond them. This opened the door for foreign militaries to take the lead in some areas. As an example, despite powerful advocacy from its military reconnaissance-strike drone crews, the US military refused to accept drones as true combat systems. Even twenty years after Roberts Ridge, reconnaissance-strike drones were still treated as intelligence-gathering aircraft that happened to carry weapons. They could not be intentionally sent on battlefield strike missions.[66] The experimental X-45 and X-47 programs demonstrated the capabilities of high-performance combat drones from 2000 through 2015 but led to no combat drone procurements.[67] In 2001, the US Congress required that by 2015 one-third of the Army's ground vehicles be unmanned.[68] That process didn't even start. Instead, in the 2020s, Western audiences marveled at the video of inexpensive precision drone strikes by Azerbaijani and Ukrainian forces in the way that world audiences had marveled at US precision strikes in Operation Desert Storm thirty years before.

THE LESSONS OF HISTORY

The history of robotic systems in combat reveals three consistent challenges. They have determined which systems make the transition from promising demonstration to real battlefield asset. They have appeared again and again across technological generations, classes of systems, and combat environments.

The first challenge is *burden*. Robotic systems that are burdensome to use or maintain often prove impractical in combat. If they distract operators, divert too many personnel, or detract from other vital combat activities, they tend to be rejected. Successful systems have used their technology to take the burden off operators and support personnel, as with "fire and forget" homing missiles, or drones that take off and land automatically. Some systems have been simple to use but require a lot of preparation before or between missions. They can be effective in offensive operations where friendly forces control the timing, but not in defense because they cannot respond quickly to unexpected events. Future systems must perform their missions with a minimum of burden on human warfighters and other resources. They must provide support, not require it.

The second challenge is *navigation and control*. Whether remote-controlled or autonomous, time and again robotic systems have struggled to handle the complexities of the real-world combat environment. Does that mound of snow conceal an obstacle? Can the vehicle make it over that barricade without getting stuck? Can the vegetation in that thicket be driven through or not? Robotic systems must address puzzles like these while also handling tactical maneuver and weapons employment. The enemy does his utmost to add additional problems. Navigation in the air is simpler; that is why aerial systems have dominated for the past hundred years, followed far behind by marine systems. Future robots, especially those that interact with the ground domain, must handle a vexing environment.

The third challenge is *vulnerability*. Robotic systems that are slow or predictable in their behavior have proven easy targets in battle. Most robotic systems have been unable to tell when they are under attack and therefore just expose themselves until they are destroyed. Until they can take autonomous evasive action, successful systems must use stealth, speed, or other factors to achieve survivability—or be attritable enough to absorb losses. Interestingly to anyone whose Wi-Fi has gone down at an inopportune time, vulnerability of wireless datalinks has been a manageable problem for purpose-built military systems in combat. Perhaps because it has been an obvious risk, forces that field remote-controlled systems have usually been better prepared to maintain their datalinks than their adversaries have been to disrupt them. However, electronic warfare capabilities are improving, so this intense preparation will have to continue wherever datalinks are required.

Historically, militaries have adopted robotic systems more quickly in combat roles, including tactical ISR, than in noncombat roles. This is somewhat counterintuitive, considering that so-called "dull and dirty" supporting roles can be less technically challenging. The potential savings in manpower and cost can often be attractive. And in fact robotic systems have become deeply embedded in a few niche roles such as target drones. However, the promise of victory in battle has been more effective in driving acceptance of robotic systems. Advantage in combat has often led to surprisingly fast replacement of earlier pre-robotic means.

Despite a hundred years of remarkable technological progress, these three classical concepts still dominate military robotics. But advancing technology has allowed them to be more fully realized. The single-use aerial torpedo and homing weapon were simpler concepts than the more general-purpose teleautomaton and had fewer challenges to overcome. They therefore reached technical maturity and acceptance earlier. A few innovations, such as loitering munitions and kamikaze or "one-way attack" drones, combine some features of the classical concepts. Many one-way attack drones resemble early aerial torpedoes, but with precision guidance and control. However, few new ideas have arisen to guide new developments, even as technology reaches the point that it's becoming practical to produce a wide variety of new robotic concepts.

Robotic weapons arose to serve a scientific approach to warfare that emphasizes certainty, efficiency, and the minimization of needless suffering. Operation Desert Storm, and many other examples, proved the potential for robotic weapons to provide superb control and deliver swift and efficient victory. The pop cultural trope of the autonomous robotic berserker set loose to slay out of control and with uncertain effects has no basis in this model of warfare. But real-world experience, from the Nazis' V-1 campaign and suppression of the Warsaw Uprising to the campaigns of ISIS, showed that robotic weapons aren't necessarily incompatible with barbarism. Nonetheless, the current surge in interest in robotic warfare is being driven by the promise of efficiency and control and the hunger for what militaries have tasted in Nagorno-Karabakh and in Ukraine: the power of universal precision.

2

ONE SHOT, ONE KILL, BY THE THOUSANDS

The Consequences of Universal Precision

Warfare has long been a wasteful human enterprise full of tragedy and loss, but in terms of physical waste and ineffectiveness, the unguided weapons that became familiar in modern industrial war have been truly astonishing. Consider, as an example, the Battle of the Somme in World War I. At the start of the battle, the British Army used more than 1,500 pieces of artillery to conduct a continuous five-day bombardment of the opposing German trenches along a fifteen-mile (twenty-five-kilometer) front. The goal was to pulverize the German positions so that waves of soldiers could rush forward and occupy the ruined trench works, killing or capturing any enemy survivors.

The bombardment maintained an overcast of smoke that blocked the sun for days. The ground shook. At night the Somme valley looked like a giant stadium filled with flickering fires from horizon to horizon as the shattered woods burned. By the time the last salvos were fired, hills of spent shell casings stood around the artillery batteries all up and down the line.

They had fired 1.73 million shells and turned the landscape into a wasteland of mud and craters.[69]

Unfortunately, artillery shells and other weapons are only truly effective when they hit something. And instead of hitting the narrow trenches and relatively small bunkers, nearly all the British shells struck only earth or trees. When the British troops charged out of their trenches, they encountered still-intact barbed wire and the fire of hundreds of undamaged German machine guns and artillery pieces. All of which targeted the now-exposed troops. By the end of that disastrous day, almost sixty thousand British soldiers lay dead or wounded.

In many ways the situation has changed remarkably little since then. Militaries still expend astonishing quantities of unguided munitions for minimal effect. The cratered battlefields and ruined cities in Ukraine, Syria, and Iraq are testament to this continuing waste and ineffectiveness—and the widespread collateral damage that comes with it.

However, this is beginning to change, and quickly. The first wave of the robotic revolution is underway: smart, precision-guided weapons are proliferating into every corner of war. The big cruise missiles and laser-guided smart bombs that revolutionized air campaigns in operation Desert Storm and thereafter were only a prelude. Today, precision is rapidly migrating to smaller, cheaper, and more plentiful classes of weapons and may soon be practically universal. The idea of "one shot, one kill" will become the standard for almost every type of weapon, large and small. By understanding the consequences of universal precision, we can see how this first wave of the robotic revolution will cause all the changes that follow.

When one missile, shell, or bullet produces the intended effect that previously required hundreds or thousands, weapon lethality increases by a hundred or even a thousand times. Such a huge increase not only offers tremendous advantages in combat, it alters the power relationship between weapons and targets and the fundamental dynamics of battle. The battlefield becomes a vastly more lethal place. The proliferation of precision robotic weapons will have major consequences for the shape of future forces, the tempo of battle, the role of information, and the need for combat AI. This first wave of robotic change, already rising, will drive and shape the subsequent waves, because the traditional military tactics and systems that worked in the past cannot survive on a battlefield ruled by universal precision.

HOW BAD IS BAD?

Just how wasteful and inefficient have unguided weapons been? The Battle of the Somme was grimly typical. An analysis after World War I found that up to 100 artillery shells, or 5,000 rounds of machine gun or small arms ammunition, had been fired during the war to produce each killed or wounded enemy soldier, known as a casualty.[70] That was not just a feature of trench warfare. Similar figures resulted from studies of the more mobile fighting in later wars. For example, a US Army–sponsored analysis of the Anzio campaign in World War II found that on average it took 200 to 225 rounds of artillery or mortar fire, or 11,000 to 18,000 rounds of small arms ammunition, to produce each enemy casualty.[71]

Such inefficiency meant, for instance, that a single US corps in World War II, of which there were dozens, expended roughly 23,000 artillery rounds per day.[72] This in turn required vast amounts of war production. The US alone produced 41 billion rounds of ammunition during the course of World War II—theoretically enough to shoot every person on Earth fifteen times over.[73] In Vietnam, US aircraft dropped an average of 70 tons of unguided bombs for every square mile of the country, equivalent to 500 pounds for each Vietnamese man, woman, and child.[74] They left approximately 20 million bomb craters, but the Americans still failed to defeat North Vietnam and the Viet Cong.

This waste and inefficiency was caused by poor precision. Directly hitting a desired target, even a big one, usually took vast numbers of unguided shells or bombs. The situation endured across land, sea, and air. In the Battle of Jutland, the great battleship brawl of World War I, fewer than one battleship gun shell in forty hit the enemy.[75] In World War II so-called "precision" daylight strategic bombing, the average bomb missed the target by more than 3,000 feet.[76] Only one bomb in five fell within 1,000 feet of its target.[77] Getting a direct hit on a target the size of a house statistically required 9,000 bombs.[78] In combat, that meant that when 108 B-17s dropped 648 bombs on a Nazi power plant, they only scored two hits, and when B-29s dropped 376 bombs on a Japanese factory, they got just a single hit.[79] The ability to hit specific targets was so poor that bombers generally resorted to simply carpet-bombing whole districts in order to hit something. To be fair, the anti-aircraft fire going the opposite way was just as ineffective. German flak gunners calculated that on average it took about 3,000 anti-aircraft artillery shells to down a single B-17.[80]

Unguided weapons today are hardly better. For instance, the average miss distance for a modern unguided American 155 mm howitzer shell at 30 kilometers is still 800 feet (260 meters).[81] That means unguided weapons continue to guzzle ammunition for little effect. In six months of fighting, the Ukrainian Army required approximately a million shells just for the 150 or so howitzers supplied by the US, a small fraction of Ukraine's total artillery.[82] Russian unguided artillery in Ukraine consumed up to 60,000 shells and rockets per day, ten million in a year, for minimal gains.[83] The cratered landscapes and static front lines that resulted resembled those of World War I in appearance and as symbols of futility.

The waste has been aggravated by tactics that take this poor precision for granted and do not even attempt to hit specific targets. They expend ammunition like water from a fire hose. Indirect artillery fire tactics often direct large volumes of shells or rockets against geographic areas to pummel any enemy forces that might be there, sight unseen. Suppressive fire tactics spray an area with rifle, machine gun, or artillery rounds to pin down enemy forces, preventing them from attacking or moving while friendly forces maneuver. Those kinds of tactics have their uses. More importantly, they do not require the ability to find a specific target, much less the ability to hit it. They made sense in an era when enemy forces could not be seen or precisely hit from a distance, and they help explain the astonishing amounts of ammunition wasted.

That staggering waste comes with a high cost, though, and not just economically. Time after time, indiscriminate firepower has proven poor at producing victory, despite causing lots of random destruction. It is expensive politically and morally, as well as militarily. All those bombs and shells that miss military targets fall somewhere, often causing collateral damage and civilian casualties, particularly during urban fighting. Indiscriminate force often hardens resistance against the countries that use it. The effect is so great that modern insurgents often strive to provoke indiscriminate retaliation, because it helps them to win the war politically.

PRECISION AND ACCURACY

What is precision? The way the term is commonly used tends to mix the related concepts of precision and accuracy. *Precision* technically refers to the ability to place weapon strikes, such as shots from a gun, close to a single point. When high precision is present, repeated shots will have little "scatter" and will cluster tightly within a small radius about their mean. One common measure of precision is circular error probable, or CEP, which is the radius of a circle within which half of the shots will land. Weapons that have smaller CEP have higher precision.

Accuracy is, technically speaking, the ability to align the center point of the weapon shots at the desired aimpoint. Imagine a target at a shooting range. A rifle with misadjusted sights could place many shots close together on a target, but all of them might be well off the bullseye. In that case, the weapon would be precise, though not accurate. A shotgun blast with birdshot could be centered on the bullseye of the target even though the pattern of impacts could be widely scattered. In that case, the weapon would be accurate though not precise. To place many shots tightly within the bullseye, a weapon must be both precise and accurate.

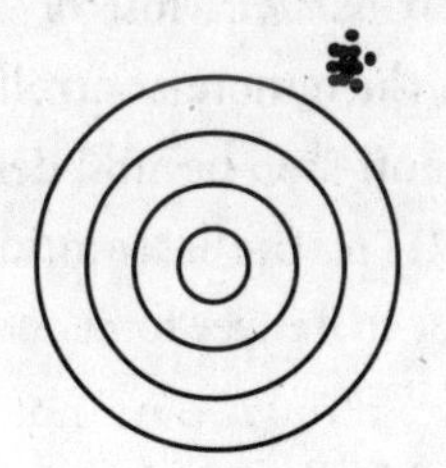

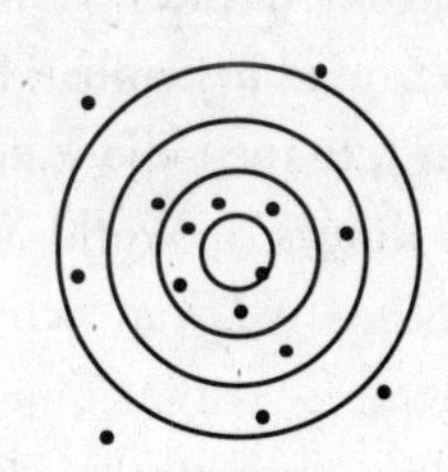

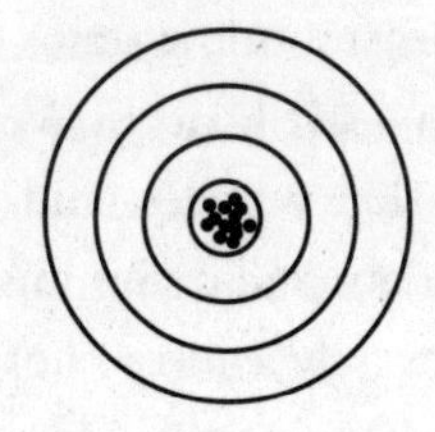

Precision vs. accuracy. Left: precise but not accurate; center: accurate but not precise; right: both precise and accurate.

The goal of precision strikes is really to achieve both precision and accuracy. Together they imply hitting the intended target while also not hitting anything else.[84] That broader definition encompasses the performance of the weapon plus the entire targeting process. Strikes that precisely hit the wrong target caused repeated tragedies and setbacks during the War on Terrorism. Similarly, powerful weapons that hit a target precisely can spread damage inaccurately over a wide area, causing lots of collateral damage. That has increased the demand for weapons that pair high precision with small radius of effect. As an extreme and somewhat lurid example, some ultra-precise missiles used in US drone strikes against individual terrorist leaders have omitted a warhead altogether, instead using a set of blades to destroy the target by contact.[85]

THE PRECISION IMPERATIVE

Military forces that used precision robotic weapons on a large scale, such as in Operation Desert Storm and in Nagorno-Karabakh, saw their military, political, and ethical outcomes improve dramatically. The major battles ended quickly and decisively, with greatly reduced civilian casualties and damage to civilian infrastructure.

Recall that the inventors of the first robots were motivated by reducing waste and suffering in war, and they identified smarter military weapons as the most obvious area of application for robotics. When Tesla unveiled his radio-controlled teleautomaton in 1898, and when Miessner and Hammond presented their homing electric dog in 1912, they touted their inventions' potential to deliver precision and certainty in battle. Most of the early robotic systems that were used in combat, from the remote-controlled explosive boats of World War I, to the radio-guided anti-ship bombs, demolition vehicles, and assault drones of World War II, to the laser-guided bombs and cruise missiles of the Cold War, were built to deliver force more precisely against chosen targets.

Meanwhile, doctrinal ideas emerged that showed how precision could win wars. In 1936, a flood in Pittsburgh destroyed the only US factory that made specialized springs used in airplane engines. The loss of that one factory halted all American aircraft production for months.[86] Thinkers at the US Air Corps Tactical School proposed that if similar critical facilities could be identified and destroyed on purpose, it could be possible to destroy an enemy country's ability to wage war without widespread destruction and without even fighting the country's army. American military leaders designed the precision daylight bombing campaigns of World War II to use that theory against the Axis powers by targeting ball-bearing plants and other industrial "choke points." Unfortunately, the poor precision of unguided bombs made it impossible to hit those targets reliably.

A few decades of technological improvement later, the air campaign of Desert Storm in 1991 finally demonstrated the power of precision air attack on a large scale. Following Desert Storm, US military thinkers saw the possibilities for extending precision to smaller targets and every level of war. As an influential airpower treatise from 1995 put it:

> One could argue that all targets are precision targets–even individual tanks, artillery pieces, or infantrymen. There is no logical reason why bullets or bombs should be wasted on empty air or dirt. Ideally, every shot fired should find its mark.[87]

EVERY WEAPON A PRECISION WEAPON

Large precision-guided aerial weapons, like laser-guided bombs, revolutionized air campaigns starting with Desert Storm. But those weapons, weighing 500–2,000 pounds (225–900 kilograms) or more, are dinosaurs today.

More advanced precision guidance capabilities are now available in much smaller packages. The Javelin anti-tank missile of Ukrainian war fame, one of several missiles that have given individual infantry soldiers the ability to destroy main battle tanks with high certainty, weighs only 33 pounds (15 kilograms). It was first produced in the late 1990s. It has been joined by more recent and smaller shoulder-launched, fire-and-forget guided missiles like the Spike SR, which arrived in 2012 and weighs 22 pounds (10 kilograms), and the Enforcer, introduced in 2019 and weighing 15 pounds (7 kilograms).

How can small weapons destroy heavily armored targets like tanks? Advances in warhead design have greatly improved armor-piercing capabilities. For instance, shaped-charge warheads focus the energy of an explosion on a single spot, like a lens focuses light to a point. Explosively formed projectiles allow this focused power to transmit over several meters. Weapons with these and other advanced warheads can penetrate surprisingly thick armor. In addition, anti-tank munitions often strike from the top, where armor is thin. Some modern anti-tank missiles fly over a tank and then explode, sending an explosively formed projectile through the thinner armor atop the turret or the hull. As a result of these advances, armor has become less protective. Relatively small anti-tank weapons can turn even heavy armored vehicles into burning wrecks.

Precision guidance can also come in a very small package. Microfabrication allows what were once expensive and complex parts, like gyroscopes for autopilots and inertial guidance systems, to be built into microchips that cost only a few dollars. Massive investment from the mobile device industry means that tiny high-definition video cameras and other components are also commercially available at low prices.

Those components have enabled a boom in very small precision-guided missiles and glide bombs. The Turkish MAM-C laser-guided bomb used on the Bayraktar TB2 drone weighs only 13 pounds (6 kilograms). Other precision bombs, such as the Pyros, Fury, Saber, and Hatchet, are as small or smaller. Armed first-person view (FPV) quadrotor drones can provide high-quality color video, high remote-control maneuverability, and anti-tank explosive power in a package under 7 pounds (3 kilograms). They cost under $1,000, less than 1 percent the cost of a Javelin. The Switchblade 300 loitering munition weighs only 5.5 pounds (2.5 kilograms) and can be toted in a backpack. Such weapons are being built in more countries, and they will become ever more capable and less expensive.

Some artillery is also becoming precision guided. The US Army fielded the Excalibur artillery round starting in 2007. It uses GPS and inertial guidance to give artillery a CEP of only six feet (two meters).[88] While the US Army expected to use them in small numbers for special purposes, they became mainstays of the Ukrainian defense against Russian forces, until Russian forces learned how to jam the Excalibur's GPS guidance. New methods allow for more modern precision guidance packages to be added to ordinary unguided artillery shells at low cost. The demand for more guided shells is intense.

Precision guidance is coming even to small arms like infantry rifles. Putting active guidance into each rifle round is unnecessary. For line-of-sight weapons like rifles, miniature computerized fire control systems similar to those on battle tanks are being built right into the gunsights, automatically compensating for factors like target range and movement, bullet drop, and crosswinds. A smart gunsight system fielded in Israel allowed untrained recruits to hit moving targets with their first shot 70 percent of the time, equaling the performance of top marksmen.[89] The Israeli Defense Forces procured it in the thousands for use by Israeli soldiers, and the technology is being applied to remotely operated robotic gun stations.[90]

The specific examples will doubtless evolve in the coming years. The numbers and varieties are increasing as technology advances and costs decline. However, the trend is clear. The technology to give every sort of weapon, from small drones to artillery to shoulder-fired missiles to infantry rifles, single-shot precision is advancing and spreading across the globe. They are all alternative means to accomplish the same goal: putting munitions precisely on enemy targets.

A REVOLUTION IN LETHALITY

The transition to single-shot precision will increase the lethality of most weapons by a factor of a hundred to a thousand. Most of our contemporary military forces and tactics, such as the kinds of vehicles and vessels we use, how our units are organized, and how they fight, were developed before this transition began. How significant will the consequences of this transition be?

In 1964, military historian Trevor N. Dupuy introduced the concept of weapon lethality as a means for analyzing the effects of advancing weapon technologies through history. At the time, most military thinkers measured weapons by firepower, which was their output in terms such as rounds per minute, or by the throw weight of artillery shells per hour. Instead, Dupuy focused on their effects on the enemy. He defined weapon lethality as "the inherent capability of a given weapon to kill personnel or make materiel ineffective in a given period of time."[91] He proposed a universal "lethality index" that allowed weapons from different periods to be compared against each other. The smoothbore musket of the Napoleonic period received a lethality index score of 47. The late 1800s breech-loading rifle received a 229. The World War II–era machine gun received a lethality score of 17,980, due to its high rate of fire. The World War II 155 mm howitzer scored approximately half a million.[92]

According to Dupuy's index, a World War II–era machine gun was 382 times as lethal as a Napoleonic musket, and the World War II 155 mm howitzer about 125 times as lethal as a Napoleonic field gun. There is no doubt that a Napoleonic regiment would have been cut down in minutes on a World War II battlefield. Due to the tremendous increase in weapon lethality, the shape of military units and their tactics had to change dramatically between those periods. In particular, Dupuy noted that greater weapon lethality forced greater dispersion of military formations. He even proposed a mathematical relationship between weapon lethality and dispersion.[93] World War II forces fought in much more dispersed formations, used low-visibility colors and camouflage, and emphasized mobility.

In the past, short ranges and low weapon lethality required the massing of forces. In the era of robotic weapons, long effective ranges and high lethality replace mass with "effective mass," which is the massing of effects. The tightly packed formations of the Napoleonic period helped units to mass the firepower of their low-lethality muskets. However, those formations would be serious liabilities when facing the more lethal weapons of World War II.

Dupuy didn't anticipate modern precision-guided weapons. He assumed that accuracy or precision was always about the same. As we have seen, a further increase in lethality of one hundred to one thousand times is reasonable in going from unguided munitions to "one shot, one kill" precision. That's similar to the increase between the Napoleonic Wars and World War II. We can expect similarly dramatic changes to forces and tactics as a result. We can expect that a present-day force such as an armored battalion would be cut down in minutes on a future robotic battlefield. Many other features of today's forces that made sense in the past may also become liabilities in the age of universal precision.

WEAPON-TARGET ASYMMETRY

Size itself may become a liability. The increasing lethality of smaller weapons due to precision is breaking the centuries-old symmetry between weapons and targets. Since the days when individual soldiers faced each other with hand weapons, a weapon system could only be reliably defeated by a weapon system of at least similar size. For example, small warships fared poorly against bigger ones. The only gun capable of penetrating a battleship's armor was the heavy battleship gun, which required another battleship to carry it. Similarly, it was a truism of armored warfare that the best anti-tank weapon was another tank.[94] As bigger tanks emerged that carried thicker armor, their opponents needed bigger tanks to carry the heavier guns needed to penetrate that armor. Engineers strove to build weapons that were ever bigger and more powerful. In the aggregate, this meant that opposing forces tended to be symmetric with each other. If a fleet had a dozen battleships, an enemy fleet seeking to defeat it needed a similar number of battleships, and so on. Military leaders and statesmen compared the numbers of battleships, tanks, aircraft, and soldiers that they possessed to those of their allies and adversaries to assess the balance of power.

This symmetry held across different weapon types because of poor precision. For instance, an anti-aircraft shell could theoretically bring down a large bomber. However, the shell had to be fired from a large anti-aircraft gun, and due to poor precision thousands of shells had to be fired to shoot down the bomber. Defeating a bomber reliably required a combination of anti-aircraft guns and shells that was similar in magnitude to the bomber. In fact, analysts in World War II calculated that the average cost for German anti-aircraft gunners to bring down a heavy bomber was $106,976, which was comparable to the cost of a B-17 bomber at the time.[95]

When General Billy Mitchell demonstrated in 1921 that early bomber aircraft could sink a battleship, US senator William Borah asked, "If a thirty thousand dollar airplane can sink a forty million dollar battleship," why build battleships?[96] The effects of poor precision made that idea premature, but it started to become real in World War II when the first precision-guided weapons such as Germany's Fritz X really did allow single bombers to cripple large warships under wartime conditions. Battleships largely disappeared after the war. Some observers are asking today: If a Javelin missile or an even cheaper armed drone can destroy a multimillion-dollar tank, why build tanks?

Today, a single F-35 fighter costs over $80 million and requires over 40,000 man-hours of labor to build.[97] Yet a small robotic weapon that can destroy it or another advanced warplane, particularly when the plane is sitting on the ground, costs a tiny fraction. As part of a US Naval Postgraduate School project back in 2006, students used a radio-controlled plane to build a simple remote-controlled "aerial IED" capable of attacking parked aircraft. As they reported, "not including the cost of an explosive payload, the midshipmen were able to build this aircraft for a little under $300. Imagine a terrorist or insurgent group trading a $300 guided aerial IED for a $200 million C-17."[98] Recent Ukrainian attacks have destroyed Russian bombers and transports using that very method.[99] Advances in autonomy enable those kinds of attacks in large numbers. When those kinds of exchange ratios occur due to weapon-target asymmetry, the staggering economic costs are nearly as powerful as battlefield losses in forcing change.

Weapon-target asymmetry describes the increasing ability of small, inexpensive weapons to reliably destroy large, expensive targets. It is a key consequence of the advance of precision weapons, and it's a powerful tool for predicting future changes to the character of warfare and military forces. With so much of the precision revolution still to come, this asymmetry will be an increasingly visible factor on the battlefield.

SURGICAL FIRE: 100 PERCENT HITS IS NOT THE LIMIT

When the circular error probable, or CEP, of a weapon decreases to less than the size of the target, the probability of achieving a hit on the target approaches 100 percent. However, achieving 100 percent hits is not the ultimate limit. The trend can continue much further.

As weapon precision continues to improve, it enables a weapon not only to hit the target but to hit a specific aimpoint within that target. In Operation Desert Storm, laser-guided bombs demonstrated that capability against large buildings. In one famous example, an F-117 pilot directed a laser-guided bomb into the central ventilation shaft of the Iraqi Air Defense headquarters, devastating the building with a single hit.[100] Precision munitions have often been used to strike specific parts of large structures, such as the structural supports of Vietnamese bridges or the access tunnels of al-Qaeda cave complexes. Now, next-generation anti-ship missiles are providing the capability for an operator to choose specific aimpoints within a ship.[101] That allows a relatively small missile to damage critical systems such as a ship's engine or radar. Some very precise US drone strikes have hit an individual terrorist sitting in a specific seat within a motor vehicle while sparing the other occupants of the vehicle.[102] Ukrainian drone operators have showcased their ability to drop small anti-tank grenades precisely onto the weak points of Russian armored vehicles, even into open hatches. That enables a small, cheap grenade to destroy a multimillion-dollar tank.

The capability to hit points within a target with surgical precision increases the lethality of small munitions against large targets. Even small weapons can be devastating if they are precisely directed against critical vulnerable points. There are many examples of powerful targets being disabled by "lucky hits." For example, in 1940, Britain's largest battlecruiser, HMS *Hood*, was destroyed by a single shell from the German battleship *Bismarck*. It pierced her deck in just the right spot to travel into an ammunition magazine and ignite an instantaneous secondary explosion that blew the *Hood* to pieces. Imagine that soon such "lucky hits" will not be improbable accidents but the normal result of any attack. Large targets will become only as strong as their weakest point. In the hands of the fictional assassin John Wick, even a pencil could be a lethal weapon when used to precisely strike an opponent's critical points.[103]

Surgical-level precision increases weapon-target asymmetry. It gives a potent attack capability to small platforms that might have been unable to carry effective weapons in the past, like small observation drones. It also means that a given combat platform can carry many more weapons. For instance, an aircraft that in the past might have carried four 500-pound bombs for use against armored vehicles could potentially carry up to eighty 25-pound surgical fire munitions, enabling the aircraft to disable twenty times as many vehicles during a single mission.

Today, the early examples of surgical fire attacks require manual selection of aimpoints, for instance by using a laser designator or the careful video-guided positioning of a small drone. Soon, active terminal guidance powered by AI could automate that process. Image-processing algorithms could automatically identify the type of target under attack, look up the vulnerable points associated with it, and steer the weapon into one of those vulnerable points. Hence, robotic weapons could automatically use surgical fire to ensure that every hit is a lucky hit.

THE ACCELERATION OF COMBAT

Universal precision also implies a dramatic acceleration in the speed of combat. When it takes only one shot instead of many to destroy a target, combat happens much faster. When large precision weapons were first used at scale in Desert Storm, the efficiency of precision-guided bombing meant that air forces could attack and hit many targets at the same time, resulting in shock and paralysis. As General Ronald Fogleman, the Air Force chief of staff, put it, when the transition to precision-guided attack is complete, US air forces "may be able to engage 1,500 targets in the first hour, if not the first minutes, of a conflict." The result could be a conventional attack with the speed and shock of a nuclear strike, but with much greater discrimination.[104] Those concepts became codified as a new airpower doctrine of effects-based operations, based on parallel attack.[105]

As precision guidance migrates to smaller weapons, the same dynamic of speed, shock, and paralysis will apply to tactical engagements on the ground. An increase of one hundred to one thousand times in weapon lethality due to precision may result in a similar increase in speed. Because it will only take a short time to hit every visible target, high-intensity battles or firefights may only last a few minutes, perhaps even a few seconds in many situations.

The traditional spectacles of massed forces moving into battle, such as columns of tanks or fleets of ships, will likely disappear. Instead of representing power, such displays will represent dangerous vulnerability. Visible forces may become like targets paraded in a shooting gallery. During the Russian invasion of Ukraine, Russian armored battalion tactical groups advanced in concentrated formations. Ukrainian drones monitored their approach, and they fell into ambushes by Ukrainian infantry with modest numbers of precision-guided anti-tank missiles. Stinging losses forced the armored battalions to withdraw. In the future, similar forces that so brazenly expose themselves to observation will be attacked simultaneously and wiped out in moments.

Without a dramatic change in the form of military forces, this accelerating effect may create a crushing advantage of attack over defense. Consider that in the past, the opening shots of any large campaign or small-unit firefight served to commence the hostilities, but they were unlikely to change the situation dramatically because most of the weapons that were fired would miss. In contrast, in the era of "one shot, one kill," the opening salvoes could tip a battle or the campaign decisively. An initial strike such as the Pearl Harbor attack, but using precision weapons, would be much more lethal and crippling. If the forces of one side can be targeted by the other, a surprise attack becomes a dangerous temptation. In this manner, the calculi of conventional engagements may come to resemble, in miniature, those of Cold War nuclear confrontations. To reduce the temptation to strike first lest one's own forces be wiped out, dispersion, camouflage, and other arts of concealment will be critical.

COMBAT AS A CONTEST TO FIND AND FIX THE ENEMY

On the future battlefield ruled by precision weapons, anything that can be seen can be hit and killed. Therefore, we can expect future forces to strive not to be seen, while making maximum effort to locate the enemy. Combat may change from a struggle to hit the enemy into a struggle to find and target the enemy.

A strike using a precision weapon includes a sequence of steps called a "kill chain." Most of the steps are about collecting and processing the necessary information to target the enemy. The simplest version of the kill chain is "find, fix, and finish." "Find" means detect the presence of the target, "fix" means tag it precisely with an aimpoint, and "finish" means destroy it with a weapon. More detailed versions, which specify additional steps such as "track" and "assess," have since become popular. In all cases, the actual weapon strike is just a culminating step.

The contest to find and fix the enemy will become more explicit and intense. The US Air Force and other military services have built a colossal multilayered intelligence, surveillance, and reconnaissance (ISR) information enterprise to provide the information to feed today's kill chains. It encompasses sensors ranging from small tactical drones, to powerful airborne systems like the airliner-based Rivet Joint and E-7, to constellations of surveillance satellites. The US even established a new military service, the Space Force, to operate the growing network of space systems to collect and move data. All those are backed by armies of intelligence specialists analyzing ISR data and making it useful for battlefield commanders. Data networks bring all this data together to create a real-time picture of the battlespace and to coordinate actions by friendly forces, a process sometimes called "network-centric warfare." When there are many networked sensors and weapons, they form a "kill web" that lets a kill chain be completed using any combination of those networked forces.

Targeting decisions lie at the center of network-centric warfare. If warfare was about wholesale destruction, only nuclear weapons would be valued because they accomplish that far more effectively. To the contrary, in real war, choosing targets carefully is vital, and the decisions involve a lot more than just pulling a trigger. The military understands targeting as a comprehensive process. Current US joint doctrine describes targeting as "the process of selecting and prioritizing targets and matching the appropriate response to them, taking account of command objectives, operational requirements and capabilities."[106] This is a systematic and multidisciplinary process and a command responsibility that requires a commander's oversight and involvement. The process involves different areas of expertise and internal checks, starting with intelligence gathering and including the designation of the aimpoint for the munition. It then includes the assessment of effects following the attack.[107] The responsibilities of targeting place tremendous burden on those overseeing the use of precision weapons.

AI MUST ASSUME SOME TARGETING RESPONSIBILITIES

The flood of ISR data is rapidly outstripping the capacity of human analysts to absorb it. In 2019 the US director of national intelligence stated that under current trends, American intelligence organizations will need more than eight million imagery analysts, more than five times the number of individuals that hold top-secret clearances in the entire government.[108] That's before the rise of universal precision. That burden can't be pushed onto warfighters. Modern warfighters are already saturated with demands. As history shows, successful robotic weapons use their "smarts" to take the burden off the warfighter.

The advance of AI is helping to address this barrier of complexity and burden. Analysts in intelligence centers can use AI to efficiently scan vast amounts of ISR data such as satellite imagery and high-definition video to quickly find potential targets. Warfighters and decision-makers can use AI to help analyze complex and rapidly evolving pictures of the battlespace, distinguish important changes from unimportant ones, and make faster and better-informed decisions.

Unmanned systems can use AI to do some of their own analysis and lower-level decision-making without sending burdensome raw data. After all, this is what we expect of manned systems. For instance, the crews of patrol aircraft looking for enemy vessels don't simply beam back video to headquarters for analysts to assess. They do their own assessment and send notice when they find something. Edge computing using AI will allow unmanned ISR systems to act in a similar way to build a real-time digital picture of the battlespace.

In addition, AI will enable the countless precision-guided weapons to find their targets without overburdening the human warfighters. While it might sound radical, early forms of AI have long provided capabilities that allow smart weapons to perform some targeting tasks. "Fire and forget" missiles are already common in air and naval combat. Such homing weapons must be able to distinguish their targets from background clutter or other noise. They also must reject interference from countermeasures like infrared flares or radar-reflecting chaff that is intended to confuse or spoof them. New weapons use high-definition imaging sensors and image-processing software to assess which objects in view are real and which are flares, decoys, or the results of electronic interference or background noise, and they then decide which object to pursue. They examine scenes in different parts of the electromagnetic spectrum, called "multispectral imaging," and look for distinctive shape or movement. It is only a short step from selecting the real target among fake ones to selecting the target from among other objects.

Many missiles have functionality called "lock-on after launch" that allows them to use their own seekers to acquire targets once they approach a suspected target area. This is the case, for instance, for many anti-ship missiles. Even in the 1980s, frontline anti-ship missiles like the Swedish Rb-04 were built to be launched from long range against enemy fleets—and then use their own radars and logic to identify and select which ships to attack. The two Exocet missiles that sank the British transport ship

Atlantic Conveyor during the Falkland Islands conflict were decoyed away from their first target by chaff, and then they found the *Atlantic Conveyor* by themselves. Modern anti-radiation missiles are often fired without any targets, effectively daring enemy radars to illuminate them, in which case they select the radars from among all other radio frequency transmitters and attack them autonomously.

Lock-on after launch was widely accepted based on 1980s technology. As modern AI advances, weapons will make ever-more-sophisticated kinds of target assessments. Such AI capabilities are similar to those that smartphone cameras use to automatically detect human faces or identify other kinds of objects. The precedent is well established for image-processing software and AI to help with some steps of the kill chain. As the demands increase due to universal precision, the technology will advance and provide help with more aspects of targeting.

SIGNATURE-SEEKING WEAPONS

Late in the Cold War, the US introduced weapons that could autonomously target tanks on a flat landscape. For instance, the CBU-97 Sensor Fuzed Weapon is an air-delivered bomb that releases forty submunitions that independently seek out large vehicles based on laser object detection and temperature differences.[109] While it demonstrated devastating capability against masses of armored vehicles in the open, that combat scenario rarely occurred in the subsequent decades. The weapon was poorly suited to the more complex situations of the Balkans and the post-2001 counterinsurgency wars. With the return of high-intensity warfare in places such as Ukraine, smart anti-armor weapons are coming back into demand.

However, the days of tank battalions driving into battle across open desert or grassland are fading fast. Future weapons will have to discriminate targets from nearby objects that are outwardly similar, such as military vehicles from civilian vehicles, warships from commercial vessels and boats, and even human combatants from noncombatants. That will require sensors and AI that can detect and assess a complex set of characteristics that match specific targets and distinguish them from others. In general, such a set of characteristics is called a "signature," and we can think of the weapons as signature-seeking weapons.

Imagine a near-future signature-seeking missile or "kamikaze drone" attack against combat aircraft on an airfield. A commander orders a group of missiles or drones to proceed to the airfield and destroy the enemy aircraft on the ground, the same way he or she might give that order to a squadron of attack pilots. The weapons fly to the airfield and then use multispectral imaging and AI to find objects that match the size, shape, and other characteristics of the combat aircraft being sought. The AI rejects buildings, ground vehicles, and civilian aircraft in the same area. The weapons allocate the targets among themselves and strike the combat aircraft with precision and accuracy, avoiding wasteful misses or unintended damage to other objects. Such an attack would be fast, devastating, and much more discriminate than attacking the airfield with traditional weapons like fragmentation bombs. The particular AI capability that is required in this example has already been used to automatically identify and count the different models of aircraft at airports using overhead satellite imagery.[110] Such an attack would be practical because combat aircraft have distinctive, distinguishing signatures. Once ordered, it could be carried out autonomously and with high certainty, with low burden to friendly forces and no risk to friendly aircrews.

BEYOND PRECISION TO SELECTIVITY

Precision was a useful concept throughout an era when aimpoints were manually designated and the job of a weapon was simply to hit as close to its aimpoint as possible. That era is ending. Military leaders and analysts will need a new metric and way of thinking for an era in which that kind of precision is taken for granted, when aimpoints are selected automatically and limited steps of the targeting process are conducted or assisted by signature-seeking AI. This approach should also measure complete effects-based performance that explicitly includes the ability to not hit anything except the intended target.

Physicians and scientists who develop treatments for medical conditions such as cancer deal with a very similar situation. They need treatments that destroy specific kinds of cells, such as cancerous cells, while leaving everything else, such as healthy cells, unaffected. It is impractical for physicians to manually designate each cell to target, so they rely on the treatment's own ability to find the chosen type of cells and distinguish them from all the other cells that may be present. The analogy to smart weapons is so direct that physicians and scientists often refer to such medical treatments as "smart bombs" or "magic bullets" and call their desired effect "hitting the target."

Medical scientists assess those kinds of treatments using the metric of *selectivity*. They define selectivity as "the ability of a drug to affect a particular population, i.e., gene, protein, signaling pathway, or cell, in preference to others."[111] In the military context, selectivity would describe the ability of a smart weapon to affect a chosen type of target in preference to all the other types of potential targets, objects, or features that may be present.

As described earlier, targeting is an inherent command function. A commander is responsible for ensuring the effectiveness of the entire targeting process, whether it is performed by humans or machines. That is vital for reasons not only of ethics but also of military utility, as poor targeting can deal a setback to the entire war effort.

In the future, a commander's ethical responsibility in deploying signature-seeking robotic weapons may become more like that of a physician prescribing a treatment for a cancer patient. The doctor doesn't approve the destruction of each cell targeted by a cancer therapy but still holds clear responsibility for the treatment. For example, a commander or responsible human controller might direct signature-seeking weapons into a designated area with the commander's intent to strike all the targets matching a specified signature, such as tanks or parked combat aircraft, and leave all others untouched.

The selectivity would depend on the fidelity of the signature, the capability of the chosen signature-seeking weapon, and the environment, including the other objects present. A weapon that could have high selectivity in one setting, such as attacking armored vehicles in the open desert, might have poor selectivity in another. Therefore, the commander's responsibility would be to ensure that the specified target signature is specific and detailed enough to exclude other incorrect targets that might be present, and also to ensure that the chosen weapons have the capacity to detect and assess the signature they are seeking with a high degree of confidence.

Continuing this analogy with medical science, the military concept of weapon lethality is also analogous to the medical concept of efficacy.

Efficacy refers to the power of a treatment to produce the desired effect in the targeted population. In cancer treatment, for instance, the desired effect is usually to kill the targeted cancer cells. A treatment that reliably kills the cancer cells, but that also kills many healthy cells, has high efficacy but low selectivity. A treatment that affects only cancer cells but isn't strong enough to kill them has high selectivity but low efficacy. An ideal treatment, of course, combines high efficacy and high selectivity. In a similar way, an ideal AI-enabled signature-seeking weapon would combine high lethality and high selectivity.

Selectivity modernizes and future-proofs the idea of precision, ensuring that future weapons deliver the highest levels of precision in the dawning age of AI-assisted targeting. Focusing robotics and AI toward the goal of creating signature-seeking weapons that are very selective could bring powerful advances politically and ethically as well as militarily. The high selectivity that lets a military force deliver precise effects for military efficiency also provides the discrimination needed for weapons to be lawful and ethical. Surveying the physical aftermath of recent urban battles is to witness the awful costs of poor selectivity. That includes battles in cities like Gaza, where technically "precise" but excessively large and powerful weapons were widely used. The cityscapes are often reduced to rubble and debris, with half-collapsed buildings and vast destruction of civilian utilities. The extent of the collateral damage to infrastructure suggests the likely collateral injury to noncombatants who were unable to escape the fighting and to the livelihoods ruined by the economic devastation. Whichever side has control of the area afterward faces a great burden in restoring its viability. By fully leveraging the emerging capabilities of precision robotic weapons, militaries may or may not achieve the early visionaries' optimistic dreams of eliminating warfare among men, but they could greatly reduce the tragic practice of "destroying a city in order to liberate it."[112]

3

LESSONS FROM BATTLE

Revolutionizing Combat Roles and Tactics

For those who were watching, scenes from the fighting in Nagorno-Karabakh foreshadowed what would soon be repeated on a much larger scale in Ukraine. Weeks into the onslaught, one of the last surviving Armenian surface-to-air missile systems, a modern Russian-made Tor-M2KM mobile system, made a stand near Khojavend in eastern Nagorno-Karabakh. After attempting to shoot down Azerbaijani drones, its crew stowed its radar and slipped the system into a high-roofed truck garage next to a house surrounded by pine trees to hide it from view. Unknown to them, the operators of an Azerbaijani TB2 drone watched the whole procedure through the drone's long-range camera. Circling high overhead, they marked the garage with an infrared laser spot. A few minutes later, a loitering munition cruised into the open garage door where the rear of the Tor was still visible. The explosion wrecked the rear of the Tor and partially destroyed the garage roof. Clambering over smoking rubble, one of the Tor's crew members struggled to squeeze out of the garage at the rear of the vehicle just as one of the TB2's laser-guided glide bombs hit the same spot, throwing smoke and pieces of metal in all directions. For good measure, the TB2 lased the remaining part of the garage roof for a final laser-guided missile strike that blew the rest of the burning garage and its contents to pieces.[113]

Precision strikes against ground troops came at a relentless pace. Turkish E-7 radar surveillance planes flying in nearby Turkish airspace could see the whole Nagorno-Karabakh battlefield and fed ISR data to their Azerbaijani allies.[114] Strikes came by day and night. Weary Armenians said it felt as if one day of this new war was equal to three months of the first Nagorno-Karabakh war.[115] One company of soldiers attempted to sally out from their base but found themselves under deadly attack by TB2 drones. They fled to their fortified barracks where they crowded through a pair of heavy bay doors just as a laser-guided munition streaked in among them. Elsewhere, Armenian soldiers huddled in extensive trench works. They said that whenever they heard the wail of a diving Israeli-made Harop loitering munition, they had only seven seconds to run or die.[116] "We could not hide," one survivor said, "and we could not fight back."[117] Explosions tore through trenches and the soldiers who took shelter there. Some fled into the yawning entrances of fortified dugouts. Small loitering munitions and glide bombs followed them. Others abandoned the trenches and huddled in groups in earthen revetments that had been built for the armored vehicles that had fled, or under nearby road bridges. Wherever they could be seen from the sky, the drones and their precision weapons found them and they died together.

DEATH TRAPS

US Army General George Patton said during World War II that "fixed fortifications are a monument to the stupidity of man." He saw that fortifications like those of the French Maginot Line, the German Siegfried Line, and others had ceased to be effective in an age of firepower and rapid maneuver. Worse, they had become liabilities. There was no wall that couldn't be breached by modern bombs, and the Army Air Forces, for all their imprecision, loved to saturate fortifications with bombs and pound them into rubble. They became death traps for their defenders.

With the dawn of universal precision, the same is becoming true even for the smallest fortifications. In Nagorno-Karabakh, narrow trenches became mass graves. Trenches only provide protection from the sides, from where most fire used to come. They provide no protection from top attack. When robotic precision weapons are widely available, trenches and other fortifications are just a way of gathering soldiers and equipment into easy-to-find places where they can be conveniently fixed and destroyed.

Armored vehicles resemble small fortifications, like bunkers or pillboxes. They can move, but their mobility is nothing compared to the speed of a missile. Air Force General Merrill McPeak observed in 2004, following US precision air campaigns in the Balkans and Iraq, that when US airpower threatened, adversary forces tended to abandon their tanks and flee. As he put it, "enemy soldiers facing precision air power now simply separate themselves from their equipment. One can hardly imagine a more pronounced change in tactics."[118]

MAXIM'S WARNING AND COMPLACENCY

When General McPeak noted the dramatic reaction of foreign armies to US precision airpower, he also noted that the US Army hadn't changed its tactics at all—because it had not yet had to. Twenty years later that is still largely the case, since US adversaries in the War on Terrorism possessed few precision weapons.

Similarly, in the early days of the industrialization of warfare, British colonial forces fielded some of the first machine guns, recently invented by Hiram Maxim. In colonial wars in Africa and India, machine guns let small British units shatter much larger native armies. That led to a dangerous complacency. A snippet from a well-known poem of the Victorian period described the attitude of British soldiers facing masses of hostile native fighters: "Whatever happens / we have got / the Maxim gun / and they have not."[119] The native armies quickly learned never to mass openly in battle. However, despite seeing dramatic effects on their enemies' tactics, the British Army maintained its own traditional formations and tactics, never considering that someday soon its enemies might use machine guns against it. Like other European armies, it marched into the first battles of

World War I woefully unprepared and suffered catastrophic losses to enemy machine guns and other industrial weapons.

Even more damning, European military experts had gotten a preview of how the new industrial weapons would change battle during a war between Russia and Japan in 1904 and 1905. European military observers saw the two sides take dreadful losses and confine themselves to trenches. Yet they only recommended minor changes, rationalizing that their own armies were more sophisticated or better trained. They were sure that if they ever went to war, things would be different.

The story of Maxim's gun warns us that we must anticipate the effects of emerging weapons on our own forces and adapt proactively before change is forced upon us. Militaries must plan for a future precision battlefield much more lethal than in the past, which may be forced on them by adversaries that might never have seriously threatened them before. We have had ample warning of the coming effects of universal precision by noting our adversaries' tactical response to our existing precision weapons, as well as by observing foreign conflicts like those in Nagorno-Karabakh and in Ukraine. If we do not heed this warning, a new round of surprise and catastrophe is waiting.

The fighting between Russia and Ukraine illustrates how the early stages of the arrival of universal precision are starting to change combat, and what adaptations they will drive. Unlike Azerbaijan, neither country had committed to robotic warfare prior to Russia's full-scale invasion in February 2022. Both had dabbled in robotic weapons but were still mostly equipped with ex-Soviet equipment from the Cold War. However, within a matter of weeks the war started to take on more and more robotic character. The military world watched the early stages of the onset of the precision battlefield as they emerged within the context of this single conflict. The results have been eye-opening for many, but the changes for warfare that they portend have only just begun. The following examples highlight some of these changes.

THE LAND DOMAIN: CARNAGE ON THE EMPTY BATTLEFIELD

Before Russia's full-scale invasion in February 2022, advisors had urged Ukraine to adopt a "porcupine" strategy. It held that a smaller defender like

Ukraine or Taiwan could defeat a stronger invader by abandoning attempts to match it symmetrically with conventional military equipment and instead field elusive forces armed with many small, inexpensive, precision weapons.[120] Such a force could be very resilient and inflict such damage on an invader that it would be forced to withdraw, like a predator repelled by the quills of a porcupine. Robotic weapons in the hands of defenders could make the costs of aggression so high that it prevents war.

In the first weeks of the full-scale invasion, Ukrainian forces used plentiful small anti-tank missiles, mostly donated by NATO, and their own armed drones to defeat the invading Russian mechanized columns. The Russian Army saw hundreds of its armored vehicles destroyed and withdrew under fire from the suburbs of Kyiv and Kharkiv. The burned-out Russian tank became an icon of the failed initial invasion. Afterward, Ukraine wholeheartedly embraced robotic weapons, making them a central theme of its armament strategy.

A year and a half later, precision weapons were a centerpiece of Ukraine's defense on the ground. For instance, precision artillery, small anti-tank missiles, and an increasing number of bomb-dropping and FPV drones anchored the Ukrainian defense of the eastern industrial town of Avdiivka, a suburb of the Russian-held city of Donetsk. Waves of Russian mechanized forces attempted to encircle the town for four months. Ukrainian forces largely stayed concealed in tree lines, factory buildings, and atop a mountain of industrial slag that overlooked the front line. Their small drones found and fixed targets from above, and the precision weapons devastated attacking mechanized groups.

The Ukrainian military claimed it destroyed up to 55 tanks and 120 other armored vehicles in a single day outside the town.[121] Russian soldiers reported that their assault units sustained 30 to 70 percent losses before being allowed to withdraw.[122] The snowy fields and roads around Avdiivka became congested with hundreds of burned-out Russian vehicles and the bodies of dead infantry. In the pauses between assaults, the Ukrainians used small remote-controlled ground vehicles to pull strings of anti-tank mines out onto the roads and fields to await the next Russian wave.[123] The Russian forces gradually abandoned mechanized tactics and switched to mass assaults on foot. The defenders finally withdrew in February 2024, after the Russian Army had lost around one thousand armored vehicles and suffered over forty thousand casualties.[124] The Russians had paid a staggering price for the gain of one suburb.

Similar situations prevailed all along the line of control. Unfortunately, the Ukrainian Army was also unable to make advances. The Russian artillery had smashed their attempt at a mechanized counteroffensive in the summer, and they had likewise abandoned mechanized advances in favor of infiltration with small units of foot infantry, echoing tactics from the later years of World War I. The front lines were effectively frozen in a bloody stalemate. "It's a war of armor against projectiles," said a Ukrainian drone operator. "At the moment projectiles are winning . . . Nobody really knows how to advance right now. Everything gets smashed up by drones and artillery."[125]

Dupuy's analysis had suggested that increasing weapon lethality would eventually force extreme dispersion and lead to an apparently empty battlefield.[126] The forces on that future battlefield would coordinate across distance and concentrate precision weapons fire on any visible enemy forces. The battlefield in Ukraine showed that eventuality has become reality. Maneuver in the open became almost impossible.

The unsurprising reaction to this has been a rush to field better defenses against drones and other small precision weapons to protect tanks and other large targets. Air defenses against manned aircraft have struggled against new robotic aerial weapons. Armed drones and loitering munitions mercilessly hunted and destroyed traditional air defenses in Syria and Nagorno-Karabakh and have done so in Ukraine. Some of that is because of their small size and low radar cross-section. Some of it is because surface-to-air missile systems often use signal processing techniques that filter out radar returns from slow-moving objects. That helps them discriminate aircraft from false targets like birds and ground objects. But slow-moving drones also get filtered out. Even when they can adjust their signal processing to see small unmanned aerial systems, most air defense systems carry small numbers of big and expensive missiles. Those are cost-effective against expensive aircraft but too few and too costly to use against large numbers of cheap and precise robotic weapons.

Both sides scrambled for stopgap remedies. Ukraine fielded scores of mobile units equipped with searchlights and machine guns to help counter the slow-moving Russian kamikaze drones such as the Iranian-made Shahed-136, which were little more than old-fashioned target drones filled with explosive. Russian troops covered their vehicles with improvised metal cages meant to protect them from drone-dropped grenades and FPV drones. Both sides fielded anti-drone jammers and other electronic warfare systems as quickly as they could. Nonetheless, by early 2024, small armed drones accounted for half the armored vehicles destroyed in some areas.

Meanwhile, militaries around the world rushed to field automatic active defense systems for armored vehicles and defenses customized for use against unmanned aerial systems. Many are focusing on electromagnetic jamming devices that can disable a drone's control system. Some are reintroducing radar-guided anti-aircraft gun systems. Newer technologies such as high-power microwave systems may help to zap groups of aerial drones simultaneously.

However, the vulnerability of today's small drones may not last. Many first-generation military drones are based on off-the-shelf civilian systems. They operate on known commercial frequencies and lack electronic hardening. Therefore, they can be disabled using simple means like handheld electromagnetic jammers. Many early kamikaze drones fly slowly and in clear view of ground defenses. Better short-range air defenses will appear soon, but upcoming generations of small drones and other robotic weapons will get tougher and more elusive. They will use hardened electronics and software, be less reliant on GPS, fly evasive routes, and be built more ruggedly to absorb damage. For instance, some newer FPV attack drones feature optical fiber datalinks to maintain unjammable connections with their controllers. Others use automatic target recognition to enable them to complete the final stages of their attacks despite short-range jamming.

Countering small unmanned aerial systems and precision robotic weapons is going from a secondary concern to perhaps the most important tactical problem for ground forces. Soon, a few defensive systems provided to units as afterthoughts will not be enough. Any future platform or facility that seeks to survive will need to bristle with defensive systems able to intercept many fast-moving precision weapons with minimal notice. Effective defenses will likely require a layered approach, with different systems supporting each other and covering each other's weak spots. Maintaining such an arrangement, especially on a fluid and moving battlefield, and always keeping it ready, will be a persistent challenge. Active defenses have typically worked best when they are at least a generation more advanced than the threat, and the threat is evolving very quickly.

THE BLACK SEA FLEET AND AIRPOWER: EXILE AND ONSLAUGHT

The rising dominance of precision weapons over large, expensive military targets was just as clear at sea and in the air. Ukraine had no navy ships, yet by early 2024 it had destroyed or disabled a third of Russia's powerful Black Sea Fleet entirely using robotic weapons. In April 2022 it sank the cruiser *Moskva*, flagship of the Black Sea Fleet, in a surprise nighttime attack using the combination of a TB2 drone and two Ukrainian-made Neptune anti-ship missiles.

The Ukrainian Navy designed and fielded increasing numbers of long-range explosive drone boats. By equipping them with video cameras and remotely controlling them via commercial satellite datalinks, they transformed one of the oldest robotic weapon concepts into a lethal force. Making their combat debut in October 2022, by early 2024 they were executing mass nighttime attacks against Russian warships. For instance, ten sea drones surprised and sank the missile corvette *Ivanovets* before dawn on February 1. Six drones struck the ship, and first-person video showed one of the drones entering the gaping hole at the waterline made by a preceding drone boat. Other drones recorded the ship's sinking using their infrared video cameras. Three more drone boats sank the amphibious ship *Tsezar Kunikov* two weeks later in the same fashion. Ukraine destroyed other ships in port using NATO-provided cruise missiles, such as the amphibious ship *Minsk* and the submarine *Rostov-on-Don*, both destroyed in Sevastopol's dry docks on September 13, 2023.

While many of the warships carried active defense systems that theoretically could have protected them, they almost never came into play. Active anti-missile defenses are only effective when they are alert and ready prior to the attack. A study in 1994 found that, in combat, anti-ship missiles hit warships taken by surprise with up to 68 percent certainty.[127] As in the Ukrainian attacks, long-range precision weapons often catch their targets by surprise even when they are operating in known conflict zones. That happened in the very first anti-ship missile strike, on the *Eilat* in 1967, and has featured in most missile vs. ship incidents since, including the sinkings of British warships during the Falkland Islands conflict and the Iraqi Exocet hit on the USS *Stark* in 1987.

Newer weapons are stealthier and can strike at longer ranges, making surprise more likely. The low-profile Ukrainian drone boats, such as the MAGURA V5, are almost invisible on radar and can operate at ranges of up to 500 miles (800 kilometers).[128] Most of their victims only had enough warning to spray ineffective machine-gun fire before being hit.

The end result was that warships, like military vehicles on land, were increasingly unable to operate in the open. Russia dispersed its surviving ships away from its base at Sevastopol into smaller, more distant harbors. Its naval ships had started the war as a powerful threat to Ukraine but before long were reduced to fugitives.

Helicopters and warplanes experienced a similar effect. Russian attack helicopters entered Ukraine as a fearsome force but were easy pickings for man-portable air defense systems or MANPADS, which could be hiding anywhere. They also proved vulnerable to late-model anti-tank missiles—and even to faster models of FPV drones. After dozens of losses, Russia withdrew its attack and transport helicopters from the front. Thereafter, both sides' helicopters could do little in combat but lob salvoes of unguided rockets toward enemy territory from well behind the lines and then turn away before they came under attack.

Ukraine's small air force kept its combat aircraft on the defensive inside its own territory from the start. After the first few weeks, as Ukraine's anti-aircraft missile defenses strengthened with new systems from NATO, Russia began pulling its fixed-wing aircraft from the fight as well. First, fighters like the Su-27 and Su-35 disappeared—and then fighter-bombers like the Su-34, and finally even the rugged Su-25 ground attack plane. Any mission over enemy territory was more likely to result in lost aircraft than real damage to a largely invisible enemy. The situation predicted after the Yom Kippur War in 1973 had finally happened: guided missiles had excluded warplanes from the skies in a major war.

The air war in Ukraine was almost entirely taken over by robotic weapons. Over the battlefield, drones and loitering munitions assumed airpower's tactical missions. Meanwhile, Russian bombers launched waves of long-range cruise missiles from as far back as the Caspian Sea. With limited

ISR capability, Russia mainly attacked fixed targets in Ukraine, including cities, power stations, and ports. In its hunger for more robotic weapons, the Russian military bought hundreds of Iranian kamikaze drones such as the Shahed-136 and set up local manufacturing in Russia. Late to the robotic weapons game, by the end of 2023 the country was massively increasing production.

Meanwhile, Ukraine used Western-supplied precision weapons such as HIMARS guided multiple-launch rocket systems to strike Russian fixed targets with assistance from NATO's ISR systems. Ukraine augmented Western strike weapons with new ones of its own. Long-range one-way strike drones of increasing range and precision hit Russian airfields, fuel depots, oil refineries, and military headquarters. Due to losses of aircraft on the ground, Russia moved more of its aircraft to bases farther from Ukraine, much as it had done with its navy ships. Nonetheless, Ukrainian agents used small armed drones to destroy costly bombers and transports at bases hundreds of miles behind the front.[129]

The Russo-Ukrainian war warns that the phenomenon of the empty battlefield must apply to rear areas as well, because rear-area functions like logistics are increasingly exposed to attack. US precision airpower doctrine has long targeted enemy transportation, supply, and command and control. Similar functions supporting US warfighting may now be increasingly endangered.

While US and NATO air forces and navies are higher in quality, Russia's are also highly capable, and the war in Ukraine should serve as a warning. Many of Russia's top-end ships and aircraft fell to Ukrainian robotic weapons that are within the means of many countries. Large numbers of more advanced weapons are further increasing the threat. Modern navies may be vulnerable even with multilayered active defenses like the Aegis air defense system. New naval doctrines such as Distributed Maritime Operations are driving the dispersion of US naval forces and an increased focus on long-range precision-strike capability for all ships.[130] They don't fully address the problem of the ships themselves, however. Fully embracing a distributed doctrine will require new types of surface ships that are less easily found and fixed.

The US Air Force is beginning to shift from combat air operations centered on large forward bases to more dispersed operations from a greater number of airfields, even some roadways. The Air Force's initiative, called Agile Combat Employment, is an important step forward.[131] However, other nations such as Sweden adopted similar measures decades ago in response to Cold War–era threats. Emerging robotic threats and pervasive ISR will require taking dispersion to a higher level.

THE INFORMATION DOMAIN: THE UNSEEN BATTLE

In Ukraine, above the battles and in the rear areas, a largely unseen struggle for ISR superiority raged. Scout helicopters and most other manned short-range reconnaissance systems were gone, and in their place floated an almost continuous constellation of small surveillance drones, several for each mile of front on both sides. Since detection by one of the specks in the sky usually resulted in incoming artillery fire, both sides used drone detectors, small jammers, and powerful electronic warfare systems to defeat the enemy's drones. The magnitude of the struggle quickly reached epic proportions. Three months after its invasion, the Russian military claimed to have downed 858 Ukrainian drones, mostly small civilian quadcopters disabled by jamming. Western analysts thought the number surprisingly high at the time.[132] Yet by summer 2023, Ukraine was regularly losing ten thousand drones per month.[133] That figure gives an idea of the scale and intensity of the battle for ISR advantage, and it has continued to increase.

The average lifetime for a small drone on either side was only a few sorties. Russia had positioned trailer-mounted electronic warfare systems just behind the front lines. Meanwhile Ukraine imported Western anti-drone jamming equipment and built a cottage industry of home-grown drone detectors and jammers. Soon, drone operators began to harass enemy drones in the sky. In October 2022, a Ukrainian operator rammed his tiny DJI Mavic into the rotors of a similar Russian drone, resulting in the first recorded case of a drone downing another one in combat.[134] By summer 2024, Ukraine fielded specialist units such as the "Wild Hornets" that used small and fast FPV drones to ram and destroy dozens of larger fixed-wing Russian reconnaissance drones.[135]

In bunkers and basement field headquarters, soldiers peered at computer screens, often flanked by stacks of observation drones headed to forward units. Ukrainian soldiers combined data from drones, intelligence from NATO allies, and reports from partisans and agents. They entered them into a data-fusion and decision-support system called Delta, which compiled a real-time, accessible picture of the battlespace.[136] Russian soldiers did the same, using more localized fire-control networks such as Strelets.[137] Unlike in past wars, the ISR struggle wasn't just a matter for specialists. It involved every unit and warfighter, down to the individual artillery battery or infantry platoon.

Each side continually searched for the other, in the radio spectrum as well as infrared and daylight. Radio direction finders let troops locate and target enemy units and drone operators. Poor electromagnetic hygiene could be fatal. An individual soldier making a cell phone call or posting a photo from the front lines often resulted in precision bombardment of that soldier's location. Deception became a prominent art form. Ukrainian forces placed convincing replicas of high-value hardware such as S-300 surface-to-air missile systems and HIMARS where Russian ISR systems could find them. Russian forces wasted many expensive missiles destroying wooden and inflatable dummies while the real hardware continued its work.[138]

The battlefield experience illustrated why armed drones are attractive to many users: They serve as their own ISR systems. They provide some ability to find their own targets, as well as to fix targets for attack by other precision weapons such as artillery and missiles. An operator can use the weapon's own camera to find targets, using a manual "hunt and peck" method. That allows even unsophisticated units to carry out precision attacks. However, this is inefficient and labor intensive. Militaries will need to automate that process to share data and achieve superior speed and scale.

Increasingly, *everything* will be an ISR sensor platform. Drones, loitering munitions, and smart missiles will "light up" their surroundings with sensors as they move. Future military ISR data networks will automatically collect that data and combine it with other sources. Data systems will quickly build an up-to-the-moment picture that communicates the locations of enemy forces. The same network can feed that data to combat units and weapons and coordinate parallel attacks against the enemy targets.

Future forces must make the battlefield appear empty by using advanced stealth and camouflage. Emerging technologies can help to mask various kinds of signatures. Those include infrared-masking fabrics and thermoelectrically cooled structures that appear invisible to infrared sensors.

Intricately constructed metamaterials can "cloak" objects by refracting electromagnetic energy around them, even visible light. Future decoys will have to become more sophisticated, exhibiting multispectral signatures that mimic the real targets. New companies are already starting to offer them.[139]

Future types of camouflage may include features that are specifically designed to evade AI object recognition. Researchers have demonstrated various unexpected "hacks" that wouldn't confuse humans but flummox various AI systems. For instance, specially patterned stickers on an object can jam an AI system with irrelevant data, causing it to misclassify the object. Some advanced AI object-recognition systems that can understand text have been made to misclassify an object just by writing the name of an alternative object across it.[140]

Because of its importance, the ISR network will become a tactical "center of gravity," a key objective determining the outcome of the battle. An invisible shadow battle of networks will take place in parallel with the physical battle. Each side will seek new ways to defeat the other's network and electronic systems, while rushing to deploy countermeasures and patches to defend its own. No datalink, guidance method, or weapon will be useful for long without upgrades and modifications. The struggle to degrade and sever the enemy's data networks will include physical destruction, electronic warfare, and cyberattack. Parts of the network that are on the ground, in the air, and in space will all be fair game, including data centers located far from the combat zone. Individual robotic systems must be able to function when cut off from the network. They must be network-enabled but not network-dependent. But whenever there is an opportunity to link up with other systems, either in their immediate vicinity or across the battlespace, the power of robotic weapons will multiply.

A CRISIS OF MANEUVER

All these observations from combat in the Russo-Ukrainian war are consequences of the rising first wave of the robotic revolution. Tactically, their broadest consequence is that maneuver is once again being suppressed by fires in all domains. The greatly increased lethality of weapons such as machine guns early in the industrial age shut down maneuver by preindustrial units such as horse cavalry and regiments of massed infantry.

Now, increasing lethality due to the onset of universal precision is shutting down maneuver by today's pre-robotic industrial-age units and platforms like tanks, ships, and warplanes.

"Fires" is a blanket military term for all kinds of strikes. It basically means raining down destruction onto enemy forces, usually from a distance, to destroy or weaken them, or to pin them down. "Maneuver" is movement of friendly forces. Fires can impose losses and weaken and demoralize the enemy, but maneuver is essential to battlefield victory. Only through maneuver can friendly forces effectively seize military and political objectives such as taking or liberating territory, breaking enemy control, rescuing captives, or freeing populations. In the first wave of the robotic revolution, attack using precision fires is dominating battle, and attack using maneuver is becoming impossible.

In Nagorno-Karabakh, only one side had precision fires. When fires rule the battlefield on both sides and maneuver is suppressed, fighting becomes a battle of attrition. Attrition means grinding, the way two stones grind against one another as they wear each other down. That is what happened in World War I. Without maneuver, battle becomes futile death and suffering. The military struggle devolves into a contest of sheer numbers and scale. Victory, such as it is possible, comes through utter military and economic exhaustion. Modern democratic societies have little appetite for that. Our militaries must avoid that situation.

This change isn't happening overnight. Significant Ukrainian counteroffensives were still possible in 2022 in Kharkiv and Kherson. Limited Russian gains were possible so long as the Russian military was willing to sustain horrific losses. But the battlefield in Ukraine remained well short of universal precision. Despite their disproportionate impact on the fighting, precision robotic weapons still accounted for a small part of either side's military. Nevertheless, the precision revolution is proceeding irrevocably across the world.

Militaries broke the bloody stalemate of industrial trench warfare by embracing the further inherent advantages offered by mechanization. They applied the new technologies to invent maneuver platforms like the tank and the warplane that never existed before. Those new industrial-age platforms delivered previously unattainable levels of mobility, protection, and firepower. Forward-thinking military leaders restored maneuver and secured massive advantage, not by substituting them into existing missions and warfighting methods but by using them to invent new warfighting methods that revolutionized combat.

Likewise, to escape the reactive mode and prevail in the age of universal precision, we must embrace the further tactical implications of the robotic revolution. We are past the point of asking what existing missions and platforms could be unmanned. The fact that robotic systems are unmanned is secondary. Instead, we must ask what inherent tactical advantages robotic systems can offer in combat, and how they may enable new revolutionary warfighting methods that can overcome the challenges of the precision battlefield.

THE TACTICAL ADVANTAGES OF ROBOTIC SYSTEMS

Robotic military systems promise powerful tactical advantages, but often these advantages are undefined. What are they specifically? Some of the most important inherent advantages include asymmetric lethality, extended presence and effect, speed of action, attritability, elusiveness, and persistence. Those are the very qualities that are needed to overcome the pressures of the lethal future battlefield. By embracing those areas of potential, and exploiting the new tactical possibilities that will arise from future robotic systems, proactive militaries may find a path toward future predominance. Here is a brief summary of each of those advantages.

- **Asymmetric Lethality.** As already described, robotic weapons provide new ways to put the most powerful enemies at risk. Small robotic weapons will continue to achieve increasing levels of lethality against large, expensive targets. Weapon-target asymmetry will continue to advance through mechanisms such as surgical fire.
- **Extended Presence and Effect.** Robotic ISR networks and telepresence extend the detailed awareness and reach of a unit much farther. Everything within the volume populated by its dispersed robotic systems becomes part of its region of control. Units equipped with robotic weapons can fight and win entire battles beyond visual range.

- **Speed of Action.** The digital information processing functions of robotic systems can perform a growing range of tasks much faster than humans. Robotic systems can execute the orient, observe, decide, and act process, known as the OODA loop, at superhuman speed.
- **Attritability.** Increasingly powerful robotic systems can be produced at ever-lower cost and in ever-greater quantity. With no human operators at risk, losses are acceptable. Robotic systems offer the potential to overwhelm opposition forces and complete missions that no manned system could.
- **Elusiveness.** Because they aren't built around human occupants, robotic systems can be small. For instance, the Black Hornet ISR drone, already fielded, weighs eighteen grams and can fit in the palm of a hand. Robotic systems can also move in ways manned systems cannot and assume an almost unlimited range of forms to incorporate camouflage and masking techniques. As improved countermeasures become able to find and fix small weapons, weapons that are even smaller could evade those countermeasures.
- **Persistence.** Reconnaissance-strike drones became vital because of their ability to spend many hours on station patiently seeking out and assessing targets. Long-endurance marine robots can operate continuously for up to a year without refueling. Robotic systems do not tire, and their attention never wanders.

TACTICAL IMPERATIVES FOR FUTURE ROBOTIC WARFARE

Restoring military power based on maneuver, which is vital for the US and its allies, will require new platforms that deliver quantum leaps in survivability, mobility, and lethality. The next chapter explores how the robotic revolution will make that possible. More broadly, robotic warfare will change the tactical dynamics of future battle. Many tactical changes may be unforeseen, but a few can be anticipated because they are already starting to appear. Three examples of tactical concepts that will become prominent include the primacy of initiative, the pervasiveness of manned-unmanned teaming, and the rise of swarming tactics. Let's take a look at each.

Seizing the Initiative

As we've seen, universal precision causes the acceleration of battle. It leads to a great advantage for the side that can find and fix the adversary and "shoot first." However, in Ukraine it often appeared that the defense had the advantage over the offense. That, as described earlier, is because of the growing obsolescence of the industrial maneuver platforms that both sides were using for offense. They are too conspicuous and vulnerable to maneuver on a battlefield that is increasingly ruled by modern ISR and precision robotic weapons.

Traditional passive defenses such as fortifications and armor are also obsolete. The defenders in Ukraine prevailed by using their robotic systems to find and fix the advancing enemy units and destroy them with highly lethal weapons before they could achieve contact with the defenders. They did not absorb the attacker's assault in the way that defenders traditionally did within strong fortifications. Instead, they proactively attacked the attacker. The idea of conducting aggressive counterattacks against an attacker, such as by preparing ambushes, is often called "active defense." Chinese military doctrine uses the more illustrative term "offense in defense." As seen in Ukraine, this practice on the precision battlefield goes beyond traditional counterattacks to instead target and strike the enemy before its attack can develop.

Strictly speaking, universal precision favors neither defense nor offense. Instead, it favors the side that holds the initiative. An attacker will naturally have the initiative, at least at first. A passive defense is likely doomed because it will never have the initiative and must attempt to parry or absorb lethal precision strikes until it is overwhelmed. However, on the precision battlefield, a defender can quickly become an attacker. Often, during the Russian invasion of Ukraine, a strong Russian force that thought it was advancing on the offensive was suddenly whipsawed by a flurry of Ukrainian precision-weapons strikes and found itself reeling on the defensive.

Future tactics will prioritize initiative. The side that can find and fix the enemy without being spotted itself will naturally tend to hold the initiative. As industrial-age platforms are replaced by more survivable robotic platforms, initiative will become a more even struggle. Attackers will strive to avoid any lag that could give an opponent the opportunity to seize the initiative. Defenders will prepare to seize on any such lag to attack the attacker and take the initiative. The asymmetric lethality, speed of action, and elusiveness of robotic weapons, including their ability to coordinate parallel attack, will make them essential to winning a struggle for initiative—in both fires and maneuver. The struggle for initiative may often come down to which side's robotic systems can outpace the other's.

Manned-Unmanned Teaming

Robotic systems provide inherent tactical advantages. However, battlefield experience shows they also need human controllers to direct them, get them out of jams, and take care of them during and between actions. This points to a tactical approach based on manned-unmanned teaming, with robots increasingly doing the fighting while humans increasingly stand back and take a directing and supporting role. Human warfighters are increasingly valuable for their holistic cognitive intelligence and judgment rather than their physical fighting abilities.

In a sense, modern air and naval forces have fought this way for a while. In recent practice, guided missiles, which evolved from pilotless aircraft, do much of the fighting on both offense and defense while humans direct and support. The pilot of a combat jet that carries a selection of guided missiles and air-launched decoys effectively commands a small fleet of military robots. The same is true for the crew of a ship armed with anti-ship, surface-to-air, and land-attack missiles. That reality is becoming more obvious as single-use missiles are joined by multi-use military drones and unmanned vessels. Current manned-unmanned teaming concepts involve manned aircraft directing "loyal wingman" drones and manned naval ships controlling flotillas of unmanned surface vessels.

On land, guided missiles are less universal, but manned-unmanned teaming has been central to robotic combat from the beginning. The German B IV first showed the power of manned-unmanned teaming in major ground combat. German troops leveraged their asymmetric lethality, attritability, and extended presence and effect to conduct armed reconnaissance and breach enemy defenses remotely. The operators controlled them and refueled and re-armed them to keep them in the fight. We can expect robots to increasingly move to the front, whether as "the first one through a breach" or the first wave of an air attack.

Until military robots are much more autonomous and self-supporting, the manned aspect will be essential for any complex or sustained operations. Unsupported robots end up in mishaps, disabled, or out of fuel. The fully independent "Terminator" is a long way off. However, dramatic advances in manned-unmanned teaming await as much more capable robotic systems emerge.

Swarming Tactics

The term "swarm" is often used as a noun to describe a group of many robotic systems such as drones, especially when they loosely collaborate, like a swarm of bees. Swarms are one of the few new robotic warfare ideas since the development of the classical concepts. However, in military tactics "swarm" is also a verb. It refers to attacking an adversary using many highly mobile forces or weapons that act in loose coordination.[141] By having greater mobility than the defender, a swarming force can harass the defender with strikes and then withdraw to evade the defender's efforts to counterattack. That lets the swarming force exhaust a stronger but less agile enemy and wear it down, like a pack of wolves harrying an elk.

There are two kinds of swarming: swarming by fires and swarming by force.[142] Swarming by fires essentially involves subjecting an enemy to so many simultaneous attacks by munitions that it is unable to defend against them and becomes overwhelmed. Swarming by force involves semi-autonomous fast-moving maneuver forces staging hit-and-run attacks from many directions. In both cases, the tactic centers on hitting the enemy force unpredictably, coordinating attacks for maximum effect, and evading the enemy force's attempts to counterattack.

Intense swarming has not figured prominently in recent wars because there has not been a dramatic difference in mobility between attacker and defender. However, robotic weapons' inherent potential for speed of action, elusiveness, and persistence, plus the ability of small agile systems to hit hard using their asymmetric lethality, is leading to a revival of interest in swarming tactics. Popular culture has embraced the idea of a swarm of kamikaze drones using a swarming-by-fires approach to overwhelm a target. Swarming by force may become much more prominent as robotic maneuver platforms and autonomy improve. Future robotic armies equipped with fast-moving robotic maneuver platforms may harass and defeat heavier and more powerful enemy forces, leaving them to curse powerlessly as they succumb, surrounded by enemies just beyond their reach.

DEFENSES OF THE NEW CAPITAL, THE FUTURE, 0530 HOURS

The night sky slowly lightens to purple above the forest while the commander huddles in his bunker, lit by glowing video screens. The shelter is shielded by an expensive, actively cooled multispectral camouflage net draped over the entrance. In the dark trenches on both sides, his fighters wait, many catching some final sleep before the first day of Year One. In the town a few kilometers to the rear, the movement's Great Leader has given the word that on this day they will liquidate all the remaining prisoners, the leaders of the old elected government and their families, and all the citizens who had supported them. Thus cleansed, the town will become the capital of a new nation, ready to set the world afire.

Their enemy, the peacekeeping soldiers of the region's bloc of nations, has attempted to reach the future capital several times in the past weeks. Each time, the movement's drones noticed and devastated them with precision attacks long before they even reached the trenches. The movement's drones and network are the best available on the international arms market. And their anti-drone defenses, also far better than anything the poorly equipped peacekeepers have, have destroyed nearly all the enemy's drones, rendering them blind. Far to the south in the so-called national capital, a special operations battalion sent by the Americans sits useless, stymied as the diplomats of the old government, the regional bloc, the United Nations, and the American government squabble endlessly about what to do.

From the sky overhead comes a strange rushing sound. Seeing nothing on the video screens, the commander steps outside the bunker and peers up in time to see neat rows of small black objects swooping down, just visible against the indigo sky. The men around him begin to shout. Gunfire erupts. The compound's automatic anti-drone defenses disintegrate several of the airborne objects, but their neighbors smoothly fill the gaps as they fall into neat lines.

Then, with a sound like a shelf full of heavy books tumbling to the floor, each row of the strange weapons strikes, all along the trenches. Strings of explosions throw fire, dirt, and sometimes men into the air. Row after row they come, tracing out the trench lines.

With momentary relief the commander sees that the thunder of explosions has already roused his quick-reaction force. The headlights of their dozen fighting vehicles are moving forward to counterattack.

Formations of black objects screech overhead toward the quick-reaction force. In a matter of seconds all their vehicles are burning, illuminating the trees with red fire.

Bewildered, the commander sprints back into the bunker and scrutinizes his screens. What is happening? Why hadn't the network given warning? Most of the monitor screens are dead now. Only the feed from a single picket drone, operating just ahead in line of sight, is still live. Its infrared cameras usually show any force coming at night, clearly illuminated in blooms of white infrared heat in the darkness, revealing the shapes of advancing soldiers and the glowing rectangles of their vehicles. The view pans left and right. Here and there, only a fleeting hint of white electronic mist is visible, like puffs of breath on a cold day, like half-seen ghosts, quickly disappearing, yet all moving toward him.

Then the feed goes dead.

Outside again, he sees his surviving men shouting and fleeing to the rear past the burning vehicles. He realizes with horror that he had not even had time to radio that his sector was under attack, and already the battle is over. The approach to the town is undefended now. The field radio he snatches crackles uselessly in his hand, dead. The fires light the forest, and a lone moth that circles overhead. Except it's not a moth. It stops in midair and fixes him in the glint of its tiny glass eye.

These are not weapons that could be bought on the international market. And the attackers know how to use them. It could only mean one thing: the Americans have come.

4

SHOCK OF THE NEW

The Evolution of Future Robotic Platforms

We've seen how the first wave of the robotic revolution is filling the battlespace with lethal precision weapons and ubiquitous sensors that are making maneuver under fire all but impossible. The familiar military platforms of the past are becoming obsolete. However, the second wave of the robotic revolution will restore maneuver through the rise of new robotic military platforms.

What will those future robotic platforms be like? When artificial intelligence famously beat human players in board games like chess and Go, it often did so by discovering unconventional new strategies that humans had never considered. The same thing happened when AI was first applied to designing a future military force. The results foreshadowed the shape of the robotic systems that will rule in the age of robotic warfare.

In 1981, Doug Lenat was a new twenty-nine-year-old assistant professor at Stanford University researching artificial intelligence. He built what was then a state-of-the-art AI system called "Eurisko" that ran on a cluster of powerful computers inside the Xerox Palo Alto Research Center. He decided to test it by applying it to a nearby amateur war-gaming competition. The object of the war game, called "Traveller TCS," was to use a fixed budget to build the most effective fleet of ships, which would then battle against other players' fleets to determine the champion. Lenat would use Eurisko to design the perfect battle fleet, which was the main challenge of the game.

He called Eurisko a discovery engine. It could accept a very complex set of rules and logical relationships that governed a problem and help discover new optimal solutions. He fed Eurisko all the war-game rules and asked it to design the optimum fleet—the perfectly balanced combination of ship types and numbers, large and small, that would defeat all others.

Eurisko's unexpected solution was to forego every kind of ship except small missile craft. Missile craft were cheap but their missiles were effective. However, they were also vulnerable. The fleet that Lenat entered in the war-game tournament was a vast array of missile craft. When battling an enemy fleet, Lenat's missile craft overwhelmed their opponent with a blizzard of missiles. The other fleet would destroy many of the missile craft, but it didn't matter. He had so many that he could afford to lose some. They were attritable. Lenat's fleet crushed all others and won the tournament by a wide margin.[143]

The next year, Lenat entered the tournament again. But this time the tournament organizers changed the rules. Originally, Eurisko had bought the maximum number of missile craft by not even buying engines for them. The new rules set fleet mobility as one of the victory conditions. Eurisko still bought as many missile craft as it could. At the end of a battle, if some of the missile craft were disabled and unable to move, Eurisko advised to sink them intentionally, thus increasing the fleet's average mobility. Lenat's fleet won by a landslide again. The organizers complained that he was ruining the game, and that if he returned the next year they would cancel the tournament.

Eurisko's achievement was a sensation in what was then the small field of artificial intelligence research, but it had limited impact beyond that. Then, twenty years later, something similar happened on a bigger stage. US Joint Forces Command ran a major war game and training exercise called "Millennium Challenge 2002." It pitted a so-called "blue" force consisting of a carrier battle group plus Marine amphibious landing ships against a cheaper "red" force meant to represent a fictitious adversary state in the Persian Gulf. The "red" force was designed by retired Marine Corps Lieutenant General Paul K. Van Riper. General Van Riper decided to forego a conventional navy entirely and build an unconventional force of swarms of kamikaze explosive speedboats and cheap cruise missiles. On the first day of the simulated war, Van Riper's red force attacked the approaching blue force. The barrage of cruise missiles saturated the blue force's sophisticated Aegis air defense system, and the kamikaze speedboats came racing in. The blue force destroyed many of red's missiles and speedboats, but it didn't matter. In the space of ten minutes the blue force lost nineteen ships, including an aircraft carrier, ten cruisers, and five of the six amphibious ships. Twenty thousand simulated service members were dead.[144]

Losing the war on the first day meant that the rest of the week's training activities couldn't take place. The organizers decided to restart the war game but forbade Van Riper from using his asymmetric approach, to ensure that the blue force would win. Disgusted that the military seemed to be ignoring a real lesson about complacency, Van Riper resigned from the war game.

Around 2010, the US Navy sponsored a series of analyses that looked for the optimum ship design for fleet survivability. Using a computer-aided engineering trade-space approach, the analyses discovered that a large ship with a powerful combination of active and passive defensive systems provided one good solution. However, the computer-aided analyses again found an unexpected optimum at the opposite corner of the trade space. The most survivable fleet was one composed of large numbers of small, stealthy craft that relied on elusiveness and attritability.[145]

If different simulations and exercises repeatedly showed that fleets of small, elusive vessels with potent weapons dominate conventional fleets, why didn't naval leaders take it to heart and start building fleets like that? While small, cheap ships might be attritable, US service members are not. No US leader would consider the crews of its vessels expendable or send any of the country's sons and daughters on a suicide mission. In addition, while small missile craft pack a big punch, in practice they are limited by their inability to carry the large, over-the-horizon radars they need to find targets at long range. If they cannot find the enemy, those powerful missiles aren't much use.

But now the robotic revolution is rendering those previous barriers obsolete. Small vessels no longer need human crews on board. They don't have to find their own targets—they can receive targeting information from other sensors through powerful ISR networks. The design principles that those naval examples brought to light are becoming practical—and not only at sea.

LETHAL DENSITY AND DISSOCIATION

The new platforms of the robotic era will address the pressures of the robotic precision battlefield and restore maneuver, by exploiting and embodying the inherent tactical advantages of robotic systems introduced in the previous chapter. They will not simply be unmanned versions of industrial-age platforms. In their evolution to exploit those tactical advantages, the next wave of robotics will take many unfamiliar forms. However, two pervasive themes will stand out: lethal density and dissociation.

Lethal density is the corollary, or consequence, of weapon-target asymmetry when applied to platform design. When small precision weapons can destroy large targets, platforms should consist of maximum precision weaponry and minimum target. They should pack the greatest amount of lethality into the smallest platform size. The total weapon lethality divided by the platform size is the lethal density. Robotic platforms can exhibit vastly greater lethal density than manned platforms. Maximizing lethal density enables a robotic platform to deliver the inherent tactical advantages of robotics including asymmetric lethality, elusiveness, and attritability.

Lethal density is already indirectly shaping the strategic considerations of military planners. For instance, many US naval strategists have quietly shifted their calculus for assessing the balance of naval power from counting the relative numbers or tonnage of ships to assessing the relative numbers of precision-guided missiles that can be brought to bear. More vertical launch system, or VLS, cells in the theater of battle means greater combat power. Seen this way, warships are basically a means for getting precision missiles into the theater. Therefore, packing more VLS cells onto warships can provide greater combat power. That is half the answer, as it shows evolution from mass to effective mass. However, just counting VLS cells can tempt navies to build big ships, because big ships can carry a lot of VLS cells. Unfortunately, in today's battlespace, large integrated platforms, much like fixed fortifications, are big targets into which people and large amounts of expensive equipment are gathered for convenient destruction by the enemy. So maximizing lethal density entails reducing the platform size as well. It provides maximum combat power while also maximizing survivability.

How can we reduce the size of platforms? Through removing human crew and through *dissociation*. As we saw in previous chapters, survival on a highly lethal battlefield requires dispersion. However, that does not only mean moving individual platforms farther apart. It can also be achieved by, in effect, "dispersing" the individual platforms themselves. Robotics allows us to dissociate a large, physically integrated platform into separate, networked subunits. In an age of robotics and networking, many parts and functions of a traditional military system no longer need to be physically attached to each other as a single platform that allows them all to be destroyed at once. Different sensors, weapons, and other functions can be located separately, often where they can be more effective. Dissociating them temporarily, such as during actual combat operations, can make each of them smaller and more survivable. Some functions can be dissociated permanently, by off-boarding them to other platforms or to the "cloud."

There are two dimensions for dissociating platforms: by separating functions onto different subunits and by replacing a large platform with multiple smaller elements. The first is exemplified by a platform that sends out specialized elements or otherwise divides off unmanned parts, reducing the size and complexity of each part. The second is exemplified by a swarm of similar robotic elements that replaces an integrated platform.

THE DISSOCIATION PATHWAY

In many ways, functional dissociation is a pathway, not a single step. Many existing platforms are already moving down that pathway. The first step on the path is locating certain functions such as weapons and sensors on separate modules that can be added or replaced to meet different needs. That is already commonplace for multi-role combat aircraft. They can take on different weapons for different missions: precision-guided bombs and missiles for strike missions, air-to-air missiles for air combat, anti-radiation missiles for suppressing air defenses, and so on. They can also mount external pods that provide specialized sensors and functions such as infrared targeting or electronic jamming. Mission planners can "mix and match" those modules to customize the aircraft for a particular mission.

The second step is deploying some of those modules off board, that is, apart from the platform. Deployable drones or other robotic systems can carry them forward into a position more optimal for doing their job, whether jamming or sensing or conducting an attack. They can have very high lethal density. Ultimately, this process can be taken to a more extensive third step, where most or all the modules perform their missions off board, and the central platform performs a limited role, such as supervising the dissociated parts as they conduct the mission or sustaining them between missions. What was an integrated platform instead becomes a set of dissociated modules. Separating sensors, weapons, and control functions will enable warfighters to mix and match them to tailor custom groupings for different missions, a concept that DARPA has termed "mosaic warfare."[146]

The second dimension, dissociation into elements, replaces a large integrated platform with an alternative construct consisting of an array of many similar small platforms. The elements may be alike but need not resemble the integrated platform. A historical example was replacing a large warship with a flotilla of small missile boats. A more modern example is replacing an aircraft or vehicle with a swarm of smaller drones. The result is a dissociated unit. While this sounds radical, it is an ancient concept. A regiment of infantry, for example, is a dissociated unit. It moves and fights as one unit but is composed entirely of separate elements, the soldiers. Robotics offers the potential to create many new types of dissociated units.

Functional dissociation and dissociation into elements are not mutually exclusive. We can easily imagine arrays of unmanned systems in which some elements play different roles than others. Some of the elements could be manned and have distinct command and control functions within the network. Full dissociation may not always be the answer: different steps along the dissociation pathway may be more practical to pursue at different times as robotic technology and operational concepts evolve.

A COMING DESIGN EXPLOSION

These considerations point to paths that lead beyond the "unmanned," or "horseless carriage–like" stage of robotic systems. Robotics allows a radical reconceptualization of military platforms. That is because platforms no longer need to be primarily vehicles for people.

Currently, military platforms are largely designed around the requirements of housing, maintaining, and protecting human crew. For instance, a large warship, whether a surface vessel like a frigate, or a submarine, or any other class, is primarily a vessel to take lots of humans on a long sea voyage, with warfighting attributes added where they are practical. Crews of military ships are particularly large even compared to commercial ships. As a classic text on ship design puts it, "Demands on a warship for crew accommodation are so great that they comprise a major design feature of the ship and the size of the ship is profoundly affected by its complement."[147] The design of a frigate devotes approximately a quarter of the volume of the ship explicitly to crew accommodations, and much of the rest of the ship is built around the crew's needs.[148] That includes everything from a bridge, to internal passageways, to a horizontal deck that the crew can walk on.

For armored vehicles, the impact of human crew may be even greater. Approximately 60 percent of the internal volume of a tank is crew space.[149] The engine, transmission, ammunition, and other parts are fitted into whatever space is left. The mass of the armor and the weight of the tank are mainly driven by the volume they need to enclose, which is determined by the crew space. Without a crew, a tank or other vehicle could be radically redesigned to accomplish the ground combat mission in new ways rather than to accommodate humans.

Modern combat aircraft dedicate a smaller fraction of their volume directly to crew, but nonetheless a great deal of their design is more subtly determined by their need to accommodate humans. Their overall size and layout are impacted by the need to place a cockpit in an appropriate location. The fact that military aircraft have distinct upper and lower sides is largely a reflection of the fact that humans prefer to be right side up rather than upside down. That preference also limits the combat maneuvers a warplane can perform, because humans can endure much higher positive g-forces than negative g-forces. Positive g-forces, like those experienced in a jet or roller coaster that is pulling upward into a climb, act to increase perceived weight. Negative g-forces, like those experienced when a jet pushes over into a dive or when a roller coaster goes over the top of a hill, act in the opposite direction. One "g" is equivalent to the Earth's gravity. With a G suit and good training, a pilot can endure short periods of nine positive g's with no ill effects, but even a few moments of negative three to negative four g's results in pain and injury such as bleeding in the eyes.[150] With no such concerns, robotic combat aircraft might be designed for greater symmetry in both appearance and maneuverability. They could be designed for performance and maneuvers that no piloted aircraft could match. In short, future robotic combat aircraft need not resemble familiar manned aircraft with the cockpit missing.

Early ironclads and submarines didn't seem like ships at all to the sailors of their times. In the same way, robotic military platforms may not look at all like familiar platforms once the design possibilities begin to be exploited. When the battlefield design pressure toward dissociation and maximizing lethal density meets with the new design freedom that is provided by removing human crew, we can expect to see a proliferation of new and unprecedented designs.

More than half a billion years ago, in the Cambrian geological period, nature figured out how to create more complex creatures than jellyfish. With a few new design options, such as the ability to make hard parts and joints and to arrange bodies into a front, middle, and back, nature produced an evolutionary Big Bang that resulted in thousands of new animal species with new body designs, including many creatures still common today. We may see a similar "Cambrian explosion" in military platforms in the second wave of the robotic revolution.

ROBOTIC PLATFORMS IN THE MARITIME DOMAIN

More than twenty years after Millennium Challenge 2002, simulations of maritime combat around Taiwan continue to produce similar results. When the US Navy simulated a battle against a Chinese invasion of Taiwan featuring large numbers of Chinese precision missiles, the Navy suffered badly. If the simulated US Navy was able to block the invasion, it did so at a terrible cost, typically losing two aircraft carriers and ten to twenty large surface warships in each war game.[151] One of the participants lamented that "they would trade every U.S. ship for more missiles" to strike China's fleet.[152]

The US Navy and other navies want larger numbers of platforms that provide greater lethality, are less easily found and fixed, are less expensive to procure and maintain, and are less painful to lose. The basic way to do that is to dissociate large, expensive ships into many smaller, more elusive, hard-hitting vessels with maximum lethal density. The shortcomings of individual vessels are greatly reduced if the vessels are unmanned. Unmanned vessels are not a new idea. After all, unmanned explosive boats were the first robotic weapons used in combat—in 1917. Robotic platforms can be inherently attritable, making all kinds of new tactics feasible.

The Ukrainian Navy lost all its conventional ships following Russia's full-scale invasion in 2022. However, it addressed that challenge by using long-range missiles, aerial drones, and explosive drone boats to damage and destroy the Russian Black Sea Fleet.[153] They set a strong example in the sense that they did not un-man conventional surface ships but instead used robotic systems to do things manned ships cannot. However, few of the specific concepts were terribly new. Future weapons and robotics systems could enable more advanced robotic concepts.

Unlike the Ukrainian Navy, the US and other global navies must operate far from their own shores, often crossing oceans. Small vessels suffer from short range and poor seakeeping in rough seas. The ability to move efficiently across large distances of open sea, which navies call "strategic mobility," is one of the few aspects where size is an inherent advantage due to physics. It is why container ships and oil tankers keep getting bigger. Yet dissociated units are more lethal and survivable in combat. That suggests that future naval vessels may be built to travel integrated and fight dissociated. It suggests a mothership concept, a bit like an aircraft carrier. Smaller robotic vessels, whether on the surface, under the sea, or in the air, could exhibit high lethal density and carry out combat actions at high speed and power output. They could then recover to the mothership for strategic mobility and replenishment. The mothership in turn could use the robotic units for combat while remaining concealed or farther out of danger.

The leaders of the US Navy's assault drone units in World War II proposed the concept of drone carriers in 1942. As in that proposal, today such carriers might use commercial hulls, and they could carry out a wide range of missions independently of scarce conventional aircraft carriers. They could also support major fleet operations, as the low-cost escort carriers often did during World War II. In undersea operations, small submarines, including so-called "midget" submarines, have long shown great promise but are hazardous to their crews. Motherships could operate fleets of small robotic submarines, even doing so covertly.

Maritime platforms have yet to begin moving down the dissociation pathway. Unlike multi-role combat aircraft, naval ships typically have all their sensors, weapons, and other systems permanently installed when they are built. Navies are exploring step one of the pathway by modularizing some of those components, producing ships that can be reconfigured for different missions and are therefore more flexible and cost-efficient. Robotic subsystems will make those components more "plug and play," eliminating much of the burden of using a new module. The original concept for the US Navy's littoral combat ships included some modularity, but the idea was not fully implemented, and the Navy decided to rarely attempt the onerous process to reconfigure them.[154] British and Danish navies are starting to implement modular concepts in their latest-generation vessels.

Modularity naturally leads to sending some components offboard—and not just ISR drones. In the 1960s, many small US Navy ships operated an early drone helicopter, the QH-50 Drone Anti-Submarine Helicopter, or DASH, that carried homing torpedoes. It gave them an effective anti-submarine capability and enabled them to extend their radius of presence and effect by attacking enemy submarines far away from the ship. Many deployable robotic systems with much more advanced technology will extend the reach and capabilities of naval vessels in new ways.

In further steps, the ability to locate some modules or components off board means that some of them need never be located with the ship at all. They may be "hosted" on other platforms or locations, and accessed and used when needed. For instance, a remote weapon station or set of missiles could be hosted on an otherwise unarmed vessel and operated remotely either to defend that vessel or contribute to a different operation. For a long time, it was common for transports and other noncombat ships to carry an auxiliary naval gun or anti-aircraft guns in case of need. Often those weapons were not terribly effective because the crew was not very proficient in using them. However, robotics and networking allow targeting data and control to be provided remotely. So the concept of auxiliary weapons could come back in a much more powerful way, with remotely commanded robotic subsystems delivering powerful capability on demand with little to no burden on the crew.

In nature there are many organisms that "free ride" on other organisms while the host organism is basically unaware of them. For example, barnacles travel attached to the skin of whales. The whale is not harmed by the barnacles, and they impose minimal burden on it. Biologists call this a "commensal relationship." Dissociation allows commensal weapons, sensors, and other modules to "free ride" on nontraditional platforms or in distant locations until such time as they are needed, adding lots of capability with little burden.

Low-maintenance commensal weapons could reside in unconventional places. Russia, Israel, China, and Iran have introduced long-range precision strike missiles that can be deployed in a standard-looking forty-foot shipping container and placed on commercial ships or on land.[155] They can go for up to seven years without maintenance.[156] Hiding such weapons in plain sight, among thousands of other similar-looking objects, can provide elusiveness. The US Army recently unveiled the Mark 70 Mod 1 Payload Delivery System. It houses four VLS cells in a shipping container. The US Navy deployed it on a littoral combat ship and an unmanned surface vessel to demonstrate a powerful "bolt-on" strike capability.[157] Similar concepts are emerging that use containers as high-density launchers for dozens, even hundreds, of small precision loitering munitions.[158]

The opening of the design space by removing human crew provides many opportunities for novel designs for dissociated elements and vessels. Some innovative defense firms are introducing small maritime craft that do not resemble traditional vessels at all. Companies such as Saildrone and Ocean Aero have provided the US Navy with unmanned surface vessels for persistent ISR that resemble solar-powered paddleboards with a vertical wing for a sail. Some can submerge to operate as unmanned underwater vehicles for short periods.[159] Innovative concepts such as high-speed hydrofoil attack vessels, fielded in small numbers such as the US Pegasus class in the 1970s and 1980s, could return in more advanced unmanned forms.

As another example of design freedom, eliminating crew volume, which is filled with air, enables buoyancy to be used as a tactical variable. Robotic semi-submersible vessels can move on the surface at high speed like conventional speedboats and then flood ballast tanks to lie flush with the water's surface for stealth and survivability. With no crew, the top surface of the vessel can be essentially flat, rendering the vessel invisible to radar in the flush position and offering no target. Acting like the nearly invisible sink-box blinds used by duck hunters, such vessels could carry arrays of vertical-launch missiles controlled remotely. Essentially a floating VLS array, they would exhibit very high lethal density and serve as powerful elements in a dissociated surface unit.

How far could this design freedom go? The aircraft carrier provides a useful starting point for considering the potential further-term implications of dissociation. Today the carrier is considered a platform, and the weapons fired by its aircraft are munitions. But what if the aircraft are unmanned, with the option of executing kamikaze-like attacks? Do they become munitions, which potentially fire other munitions? What if the carriers themselves are unmanned and attritable? The entire assembly ultimately becomes a hierarchy of dissociable elements with the distinction between platforms and munitions increasingly blurred, and based on design choices such as degree of attritability.

The Cambrian explosion has plenty of room to run in the maritime domain.

ROBOTIC PLATFORMS IN THE AIR DOMAIN

As in the past, the relative ease of navigation in the air domain means it is attracting much of the attention in the field of robotic warfare. Like the US Navy, the US Air Force and other leading air forces are eager to field larger numbers of platforms that are more lethal, less easily found and fixed, less expensive to procure and maintain, and less painful to lose in combat. Other lower-tier air forces are hungry to wield first-class airpower without the astronomical expense of fifth- and sixth-generation combat aircraft. This is producing a surge of enthusiasm for unmanned aircraft in mainstream airpower roles.

Military researchers and defense companies in the US and in other countries are racing to implement unmanned airpower concepts like those that were proposed in the 1960s and 1970s, but which the technology of the time could not deliver. One prominent approach involves manned-unmanned teaming between high-end combat aircraft and more numerous unmanned aircraft that will accompany them. The basic concept first appeared in the assault drone units of World War II. The pilots of the combat aircraft will direct and oversee the unmanned aircraft, like they might command a subordinate flight or squadron of manned aircraft. The less expensive unmanned "wingmen" will provide the desired numbers and attritability.

This approach marries successful uninhabited combat air vehicle, or UCAV, concepts that date from the X-45 program in the early 2000s with emerging advances in AI and human-machine interfaces. Active research and experimentation programs are working out how capable the unmanned aircraft need to be, how much autonomy they need, how they should collaborate with human pilots, and various other questions that are important to bring the concept into the mainstream. Unmanned jet aircraft have provided powerful accompaniment to manned aircraft since the first decoy systems, such as the Quail for strategic bombers in the 1950s. There is little doubt that future conflicts will see much more advanced drones accompanying high-end combat aircraft into battle.

This approach has appeal for leading air forces like that of the US, whose combat power is already built around top-end manned combat aircraft and their supporting infrastructure. Positioning robotic combat aircraft as accessories for them enables the robotic systems to fit more comfortably into existing operational concepts. If unmanned wingmen match the range, speed, and other performance specifications of the manned combat jets they accompany, their designs may resemble unmanned versions of conventional warplanes.

In contrast, some other countries clearly see robotic combat aircraft not as enhancements for, but as alternatives to, expensive high-end manned warplanes. They may accomplish many of the same tasks for a much lower price. While they will not match the performance of fifth- and sixth-generation combat aircraft at first, rapid evolution may soon enable them to offer a more compelling return on investment for an expanding set of missions.

Robotic platforms may also offer similar opportunities for leading air forces. For instance, much like the US Navy's past aversion to small missile craft, the US Air Force did not embrace light attack aircraft such as the OV-10 Bronco, the A-37 Dragonfly, or the A-29 Super Tucano, even though they proved effective and economical in lower-end conflicts such as in Vietnam and Afghanistan. Like the Navy, the Air Force does not consider its personnel attritable and is averse to putting them in lower-end platforms. Creating combat drone squadrons could allow leading air forces to address those lower-end combat missions with affordable, attritable platforms that are able to deploy independently of expensive high-end aircraft. As their capabilities rapidly grow, they will increasingly be able to augment manned combat squadrons in high-end combat. Where the manned oversight function is located, whether on the ground, in the air, or half a world away, may not be very important. That role could be handed off from place to place. Such units could collaborate with manned aircraft in the air, including being "picked up" by manned aircraft for local tactical control, but they could also operate independently. Letting robotic platforms develop as an independent airpower capability may let them develop much faster, as was the case for early mechanized and airmobile units in the Army, and can ensure that the US and its allies are not leapfrogged by a disruptive transition. It offers a new approach to airpower that could deliver lasting advantage.

The challenge of how best to employ unmanned aircraft in flight is focusing attention and resources, but it should not distract from an even greater strategic priority. That is the necessity to dissociate the air base. Military vehicles on land and ships at sea can maneuver in the battlespace and conduct many different missions for long periods, perhaps even for the duration of a conflict, without returning to a base. Aircraft, however, must return to base at the end of every sortie. Therefore, they are not fully independent maneuver platforms in the same sense. That means that, in effect, air bases, more than airplanes, are the key platforms of airpower. We understand that aircraft carriers are the key platforms of naval aviation, and the same relationship holds for air bases on land. Air bases are fixed in location, highly conspicuous, and few in number, all terrible attributes for combat platforms in the age of universal precision. Worse, the conventional physical layout of air bases, with aircraft exposed to the sky against a flat surface, is ideal for targeting by precision weapons. While technology has advanced by generations, air bases have hardly changed since the 1950s.

Air forces can take various measures to harden air bases by adding passive and active defenses. The Swedish Air Force trained to operate from specially prepared lengths of public highway to provide resilience to Cold War–era threats to its bases.[160] The US Air Force has also started training to conduct some operations from dispersed locations including highways.[161] Those measures are important. However, the fundamental characteristics of air bases that make them vulnerable to AI-enabled precision weapons are ultimately constrained by the aircraft that operate from them. Like other platforms seeking to survive on the precision battlefield, air bases must become still more dispersed, mobile, harder to find and target, and less subject to being disabled by a small number of hits, such as on a runway. Those conditions cannot be met while aircraft require long runways to take off and land. Future airpower must be increasingly independent of runways. Future combat aircraft must be capable of very short, and preferably vertical, takeoff and landing. They must also require less maintenance and ground support infrastructure and be built for stealth on the ground as well as in the air.

Making conventional high-performance warplanes that are capable of vertical takeoff and landing, or VTOL, is hard. The only recent example is the F-35B, which is designed for operation from small aircraft carriers such as Marine amphibious assault ships. However, VTOL craft are almost universal in science fiction. Flying saucers and all the imaginary fighters in the *Star Wars* universe are VTOL. The basic idea is just that appealing.

Robotic aircraft, however, have often been runway independent from the days of the pilotless bomber squadrons in the 1950s and the Firebee in the 1960s. Robotic aircraft can be smaller than manned aircraft. Smaller aircraft can more easily be launched without a runway, and precision weapons make smaller aircraft more lethal. The Cambrian explosion of robotic design options can provide new ways to achieve remarkable performance. Robotic aircraft offer the most direct path to generate airpower independent of runways, and they also offer all the inherent tactical advantages of robotic systems.

One main challenge is reducing burden on the ground. Today's unmanned aircraft, such as the MQ-9 Reaper, can require 150 to 200 human crew on the ground to keep them in sustained operations—much more than manned aircraft.[162] For example, an Army company operating MQ-1C Gray Eagle drones requires more than four times as many people as a company operating Apache helicopters.[163] This unexpected burden has caused Air Force generals to complain that their most critical staffing problem is manning their unmanned platforms.[164] Robotic combat aircraft should be designed from the start for minimal burden and operation from austere locations, with high levels of autonomy designed to minimize the need for intensive human ground support. This approach must consider the entire operational cycle including launch, recovery, replenishment, and preparation for the next mission. And it includes an emphasis on autonomous logistics, which is discussed further in the next chapter.

Providing runway-independent airpower will necessarily require decoupling robotic aircraft from manned aircraft, at least to some extent. The priority should be on creating the most effective and practical runway-independent units possible. Their combat power may not derive from matching the performance of conventional high-end warplanes platform for platform but from different sources, such as their ability to act as dissociated units exhibiting novel tactics and their staying power within the theater. Runway-independent units enable airpower to be delivered by air bases that are hard to find, distributed, mobile, and pack a lot of combat power in a small space.

The superior performance of high-end manned aircraft may become irrelevant if the bases for manned aircraft are destroyed or forced out of the theater, leaving runway-independent drone aircraft as the main generators of airpower. At first, the new aircraft may provide novel and useful forms of low-end light attack and air combat, supplementing conventional high-end platforms and serving where they cannot. Eventually, they may evolve to take a leading position in the changing new market for airpower.

IS THE SWARM THE PLATFORM OF THE FUTURE?

The previous chapter introduced swarming tactics. Most of us have heard about swarms, meaning large groups of small unmanned systems coordinating together in some fashion. They appear in movies, for example, and we assume that they have a big role in the future of robotic warfare. It is easy to see the spectacle of a drone light show and imagine that the drone swarm, consisting of hundreds, even thousands of small drones, could be the military system of the future. That is partially true, but it doesn't tell the whole story.

Like a fireworks show, what we do not see in the drone light shows are the hours of work that precede the show and that follow it. Burden is one of the key challenges with robotic weapons. The burden of all that work could make large swarms uncompetitive with other options—and likely impractical in most combat scenarios.

Many swarms as currently conceived are most usefully considered as dissociated munitions, or "munition swarms." A weapon consisting of a multitude of small elements is ideally suited to swarming by fires, especially when the individual elements can move independently and strike precisely. Leveraging weapon-target asymmetry, dissociating a munition into a swarm can more easily overwhelm or bypass defenses and converge on critical points on the target. Semi-random movement is appropriate. However, there are serious practical limitations to this concept. For instance, very small elements will inherently have limited range and endurance. Their ability to function as a swarm will be limited in time and space.

Since large weapons are generally faster and have longer range than small ones, it makes sense for them to "travel integrated" until close enough to the target, and then dissociate. Deploying the swarm from a larger host, like a projectile or missile, allows it to be employed where it can expend the drones' limited endurance to be of most benefit, rather than exhausting it in transit to the target location. This also eliminates a lot of the burden in preparing the swarm for use. As single-use munitions, they come "locked and loaded" and do not need to be regenerated.

Swarming to overwhelm and destroy a single target is a valid use when a target is strongly defended. However, disabling one target may not require that many small precision weapons. One or a few can usually do the job. The more important use for dissociated swarming munitions may be in striking dispersed targets. For example, the latest sensors on small drones can often reveal the mines in a minefield.[165] However, seeing them does not help much in neutralizing them quickly. A swarming munition could deploy a multitude of small drones over the area to precisely strike each mine, clearing the field in moments. As lethality drives greater dispersion on the "empty battlefield," the ability to precisely hit many separate targets at once may become more valuable than the ability to destroy one target.

In contrast to munitions, platforms for maneuver—"maneuver swarms"—must be enduring and reusable, even though they may be attritable and dissociated. For a reusable unit, the issue of recovery and replenishment, and all the associated burden involved in sustaining the swarm in longer-term combat operations, is unavoidable. And as the number of individual elements increases, and they become smaller and shorter in endurance, the burden problem multiplies. Therefore, a swarm of robots meant for sustained operations, such as manned-unmanned teaming operations, will need to have a more modest number of elements. Larger robotic platforms also have greater capacity to carry payloads like weapons and sensors that make them more versatile. This points toward maneuver swarms, compared to munition swarms, having smaller numbers of more capable and versatile elements.

Since ancient times, militaries have observed that in most cases individual elements or platforms, including soldiers, ships, and so on, are more effective when arranged in a formation than as an unorganized mob or swarm. Hence tactics were born. Armies that used formations and tactics routinely routed barbarians that did not. Today, the formations are more dispersed and less obvious. Nonetheless, whether it is the way soldiers advance on patrol or the way ships are arranged in a battle group, effective military units use intentional formations designed for maximum tactical effectiveness.

Groups of unmanned systems will be subject to the same reality. Hence, unless they are intentionally using swarming tactics as described in the previous chapter, they will be most effective by assuming formations that are tailored to the task they are performing. For instance, unmanned systems conducting a reconnaissance screen might advance in a line-abreast formation, with spacing optimized for their particular sensors to cover ground most efficiently without gaps. Part of the power of a dissociated unit, relative to an integrated platform, includes its ability to rapidly assume new formations as missions and circumstances change.

ROBOTIC PLATFORMS IN THE LAND DOMAIN

The dominant platforms for land combat in the second wave of the robotic revolution are unlikely to resemble unmanned tanks. Such uncrewed armored vehicles would neither address the vulnerability of today's tanks nor deliver many of the potential tactical advantages of robotic systems. Just as the platform for mechanized cavalry was not a mechanical horse, the platforms for robotic ground combat are unlikely to resemble industrial-age tanks and other familiar armored vehicles.

Neither are they likely to resemble humanoid robot soldiers. Humanoid robots also exemplify few of the potential tactical advantages of robotic systems. They tend to have similar physical limitations as human bodies, and therefore they provide little additional tactical benefit. They are also very mechanically complex, meaning they will have poor reliability and need lots of maintenance and repair. The idea of a humanoid military robot owes a lot to movies such as *The Terminator*, but recall that the Terminator was only built as a humanoid because that allowed it to pass for a human and infiltrate human bases. Likewise, humanoid robots will be useful in certain circumstances such as navigating interior spaces designed for humans, for instance, buildings that have doors and stairs. They may develop first for missions like supporting police operations, where they only need to perform for short periods, and then migrate into select combat use. In most cases, however, rather than duplicate what humans can do well, robots are best designed and used to do things that robots can do well and human soldiers cannot.

Tanks and similar platforms will not become obsolete suddenly. Instead, they are likely to follow a similar trajectory to battleships. Like battleships, they were designed to embody the industrial-age qualities of protection, firepower, and mobility. Indeed, tanks were originally conceived as land battleships. When precision aerial attack against battleships first suggested that small, precise weapons could destroy large, expensive targets, some suggested that battleships were obsolete. However, navies staved off their obsolescence by adding more and more active anti-aircraft defenses. The last battleships of World War II were practically encrusted with them. For instance, the Japanese battleship *Yamato*, in its final configuration, carried 178 anti-aircraft guns.[166] Nonetheless it was still sunk by air attack. Some battleships endured after the war but were useful in fewer and fewer circumstances. The firepower of their mighty guns was largely irrelevant because battle took place far beyond gun range. Their vulnerability became a greater and greater concern, and they relied upon more support from other systems, until they were no longer seen as cost-effective platforms. Encrusting armored vehicles with active defense systems can extend their usefulness a while longer, but their time is passing.

Even though increasingly vulnerable or ineffective, an established military platform such as the tank will persist until an alternative arrives that can better satisfy the need it addresses. The crisis of maneuver demands new alternatives.

Military robots extend the radius of presence and effect for land units of all kinds. This was foreshadowed long ago by the Borgward B IV. More agile, lethal, elusive, and attritable, robotic systems will increasingly do the fighting while manned platforms stand back and remain concealed, more like aircraft carriers than battleships. They may exercise control and act as motherships to sustain the robotic systems in the fight. The transition will be gradual, and the balance of manned-unmanned teaming will change over time as technology advances.

The land domain is extremely varied, and many unique missions will call for specialized robotic capabilities. For instance, ubiquitous ISR and precision weapons are driving enemy forces into cover and concealment. Uprooting enemy forces from deeply dug-in positions can be difficult and costly. Specialized ground robotic systems will help clear tunnels, cave complexes, and urban structures. Some other tasks, such as hauling heavy cargo, may require heavy ground vehicles.

For most land combat situations, however, ground robots face persistent challenges. The lessons of robotic combat history show that issues of navigation and control, vulnerability to enemy fire, and operational burden are particularly difficult for robotic vehicles that must move on the ground. Obtrusiveness and slow speed make ground robots easy prey on the precision battlefield. This would seem to suggest a pessimistic outlook for military robotics in the land domain. Nonetheless, ground combat is where robotics could have its most revolutionary impact, as seen in the next chapter.

5

THE NEW KEY TERRAIN

Control of the Atmospheric Littoral

Since the days when the remote-controlled Borgward B IV breached fortifications and spearheaded major assaults during World War II, robotic systems have held promise for revolutionizing land combat. Early combat experiences showed that teaming mobile attritable robots with manned vehicles and human controllers created powerful synergies on the battlefield. However, they also revealed the weaknesses that have inhibited military ground robotics ever since.

The most intractable issue has been the difficulty of moving and navigating on difficult terrain. The smaller and potentially more elusive the systems, such as the B IV's small cousin the Goliath, the more they struggle. Larger robotic platforms can sometimes manage their way across larger obstacles, but they tend to be slow and awkward compared to soldiers and manned vehicles. As a result, they are often sitting ducks on the battlefield.

The emerging precision battlefield is more lethal, fast-paced, and unforgiving. Future battlefields promise to be more physically challenging as well. The pivotal battles of recent conflicts have been urban fights for cities such as Mosul in Iraq, Raqqa in Syria, Mariupol and Bakhmut in Ukraine, and Gaza in Palestine. Almost two-thirds of the world's population lives in cities, and military strategists note that "urban operations in the twenty-first century are not just another type of operation; they will become this century's signature form of warfare."[167] The terrain of an urban battlescape can be vastly complex: a jumble of obstacles including barricades, craters, wrecked vehicles, and debris from damaged structures. Even with navigation using AI generations beyond that in current self-driving cars, or with continuous remote control by a human operator, the physical obstacles may be insurmountable by any ground robot.

THE NEW REGIME OF COMBAT: THE ATMOSPHERIC LITTORAL

Raise the plane of movement maybe thirty feet (ten meters) up, however, and the path is smooth and unobstructed in all directions. Flying drones can operate at this level but remain intimately engaged in ground combat. They are effectively ground forces, but they operate in the air with the tactical advantages of airpower. They maneuver in what is often called the "atmospheric littoral," which is the portion of the atmosphere adjacent to Earth.[168] It is sometimes called the "air littoral," or "air-ground littoral."[169] In terms of future military operations, it is where the following conditions apply:

1. Operations are conducted in the air, high enough that most ground obstacles are of no consequence and forces can move, concentrate, and disperse without hindrance, much like aircraft but on a local scale.
2. Operations are conducted low enough that the forces are in close and intimate contact with ground forces, able to directly attack enemy ground forces or support friendly ones.
3. Operations are conducted low enough that the forces can use large features such as buildings, hills, or large trees as cover and concealment when needed.

Navies use the term "littoral" to refer to the shallow so-called "green" waters near land, distinct from the deep "blue" water of big ships and the open ocean.[170] By analogy, the atmospheric littoral extends from the ground to an altitude of a few hundred feet. In shorthand, it can be thought of as "the air between the buildings." Contemporary helicopters often operate in the atmospheric littoral but tend not to remain there for long in combat because they are vulnerable to enemy fire at that altitude, and their size, which is required for carrying human pilots and passengers, prevents them from maneuvering safely or effectively between buildings and trees or along streets. Once they are high enough to be out of danger from enemy small-arms fire, they are, tactically speaking, out of the atmospheric littoral.

Small quadrotor or hexrotor drones armed with simple weapons like grenades were the first armed systems to operate continuously in this intermediate space between earth and sky, and they had immediate impact. When ISIS militants first used them in the Battle of Mosul in 2017, they were entirely new to combat. The cheap, reusable platforms observed Iraqi government troops and delivered fires on a small scale, making precision strikes like a new form of airpower. However, when opposing forces tried to destroy them, they realized that the drones were neither combat aircraft nor ground platforms. Weapons for use against ground targets were mostly useless. Many frustrated units blasted away with automatic weapons but were unable to score a hit on the small, elusive drones.[171] Anti-aircraft defenses that were designed for use against combat aircraft and helicopters were equally useless. Even though the Iraqi forces advancing through Mosul enjoyed dominating air cover courtesy of US fighter jets, the American pilots found their erstwhile air dominance was irrelevant to the new threat from small ISIS drones.[172] Their powerful radars and missiles provided no means of finding or engaging the elusive objects moving far below. The drones operated in a regime distinctly different from the higher-altitude domain of conventional airpower.[173] Unknowingly, the militants had opened the door to a new regime of combat distinct from traditional army ground maneuver and from traditional airpower, a regime open to both fires and maneuver, made possible by the rise of robotic systems.[174]

THE COMBAT DRONE

For several years after they first appeared in Mosul, small armed drones were improvised weapons made by attaching explosives to civilian hobbyist models that were intended for video photography. The explosives converted them from handy ISR platforms for observing the enemy into means of attack. This harkened back to the pilots of observation planes early in World War I who began to carry grenades in the cockpit to drop on enemy trenches as they passed overhead.

Before long, those early observation planes were replaced by warplanes purpose-built for combat. We can expect the same for armed rotary-wing drones. What would a platform purposely designed and built for maneuver combat in the atmospheric littoral be like? To be effective, it needs these features:

- Three-axis maneuverability, so it can maneuver up to an altitude of several hundred feet, move along multiple axes, or remain stationary. This effectively rules out fixed-wing aircraft.
- Small size, enabling it to maneuver effectively between buildings, trees, and other obstacles like cell towers and power lines. This effectively rules out human-piloted vehicles.
- Usable payload, large enough to carry light-infantry, man-portable, precision weapons with significant lethality.
- Advanced controls, with onboard sensors and communications able to sense its environment, report its circumstances, and interpret commands.
- Autonomy, giving it the ability to manage its own stability, navigation, and other functions without continuous human input, and capable of coordinating with other similar platforms.
- Endurance, sufficient to conduct meaningful combat operations and return to a logistics point before running out of energy. About thirty minutes may be a practical minimum, with the ability to return immediately to operations after visiting the logistics point.[175]

Such a platform would be a bit larger and more capable than a weaponized civilian hobby drone but built using the same technologies and low-cost supply chain. Multi-rotor drones were the first systems that have the characteristics needed to operate in the atmospheric littoral. We can imagine future systems that do not need rotors, just as the early airpower leaders could imagine future airplanes that did not need propellers. However, for the time being, multi-rotor drones can provide the necessary performance.

Heavy-lift drones in the 220-pound (100-kilogram) class are well within the state of the art, and drones that can lift up to 800 pounds (360 kilograms) are in advanced development for resupplying ground troops.[176]

In terms of weapons, dropping small grenades remains an option, as does flying into the target, kamikaze-style. But soon ground forces will have better defenses against drones operating in clear view. Therefore, combat drones will need more powerful and versatile precision weapons that can strike targets from a distance. A light machine gun with precision fire control and ammunition weighs around 10 kilograms (22 pounds), and modern shoulder-fired anti-tank missile systems weigh about the same. A weapons loadout of a machine gun and four missiles would weigh around 50 kilograms, well within the capacity of a drone.

This is similar to the armament carried on light tactical vehicles that are used by infantry and special operations forces. However, the combat drone has only about one-tenth the mass. It can also cost half as much, and if commercial supply chains and mass production are leveraged, it can be much cheaper still. That means the drone has about ten times the lethal density and provides at least twenty times the lethality per dollar. Furthermore, it does not need a crew on board to function. Then add its vast superiority in mobility, terrain independence, organic ISR capabilities, overhead vantage point, and dominant tactical position, and it offers revolutionary opportunities in ground combat. For ages, land forces have sought to occupy the high ground. Platforms in the atmospheric littoral carry the high ground with them.

Many modern military vehicles have remote-controlled weapon stations that contain cameras and sensors plus weapons. They are mounted on the roofs of the vehicles or in distant locations such as observation towers. The combat drone is like a remote weapon station that flies. It can move quickly over any kind of terrain and move to whatever position is most useful. It can also land, letting it occupy or hold a position or conduct long-term surveillance from a remote location.

Such drones can be operated by infantry units, armored units, special operations forces, base security forces, or almost any type of unit that wants to embrace the tactical potential of robotics. They can serve in manned-unmanned teaming arrangements with conventional units, similar to the B IVs of World War II but with much greater performance and effectiveness.

Defenses against drones are rapidly improving, but that will serve to accelerate the development of true combat drones. Whether the anti-drone defenses use guns, lasers, or high-power microwaves, nearly all require a line of sight to the target. They will mostly threaten early-generation drones that operate at moderate altitudes and in clear view. They will also threaten large manned platforms like helicopters, making them even more vulnerable. However, true combat drones will more aggressively exploit the advantages of the atmospheric littoral regime where they thrive. They will spend more time at ground-hugging low altitudes where they cannot easily be seen and exploit the cover of trees, buildings, and terrain.

THE DRONE ARRAY

The combat drone is mobile, lethal, and useful for a great many things. However, it is not the true maneuver platform of robotic ground combat. That appears when a group of such drones is combined into a powerful and flexible maneuver swarm. The combination merges the power of all the drones into a whole that is greater than the sum of its parts. It multiplies the power of the combat drone through the additional advantages of a dissociated unit.

As described in the previous chapter, a maneuver swarm differs from a munition swarm in that it includes fewer, more capable elements and a higher degree of tactical order and control. The term "drone array" makes that distinction clear and intuitive, and it helps avoid confusion with other ways of using aerial drones. The drone array maneuvers as a unit and is controlled as a unit, while the individual drones handle the details of moving and maintaining their formation autonomously. The array can assume different formations that maximize its advantages in changing circumstances. As a whole, it provides tremendous lethality and survivability. The ability to concentrate and coordinate the precision firepower of all the drones generates very high combat power and effective mass. The dissociation of the array, with individual drones operating well apart from each other, makes it nearly impossible to destroy it with a single hit regardless of the power of the weapon used against it, and the loss of any single drone only marginally degrades the capability of the whole.

Engineers have demonstrated the ability of a drone array to autonomously manage its own formations and behave as a single unit using small commercial drones in controlled environments. That includes complex behaviors, such as quickly changing from one formation into another and moving its formations around obstacles. It includes moving a formation through a narrow constriction such as a doorway and smoothly re-forming it on the other side.[177] Recent advances in technology are allowing drone arrays to perform similar behaviors in less-controlled, outdoor environments.[178]

The ability of a drone array to maneuver in the vertical dimension, and the freedom from ground obstacles that it enjoys, provides several pervasive advantages in combat. In many ways these combine the tactical strengths of air forces and ground forces, and they permit the full exploitation of the atmospheric littoral regime. Advantages include:

- **Speed.** As with air forces, movement through the air is fast and unobstructed, so forces can be sent quickly to achieve time-sensitive objectives, like cutting off a withdrawing enemy's route of escape.
- **Concentration.** Also like air forces, the ability to fly over terrain and over ground forces enables the ground commander to concentrate combat power at the decisive time and place anywhere within the battlespace, even places far from other friendly forces.
- **Persistence.** Like ground forces, atmospheric littoral forces can remain in position and can seize and hold terrain. Operating in close contact with the ground, they can land and remain in place for long periods to physically occupy objectives and deny their use to the enemy.
- **Mass.** Unlike other military forces, they can be arranged in space arbitrarily, enabling a unique concentration of firepower. For instance, arraying forces vertically could enable a range of drones at different altitudes to all fire at the same target at the same moment.

While the atmospheric littoral exploits many of the advantages of airpower, it is intimately connected to the fight on the ground. Drone arrays may be naturally controlled by a ground commander as part of a single integrated combat arms force. The discipline and control embodied in the drone array allow atmospheric littoral forces to coordinate closely and continuously with ground units.[179] However, they also enjoy the freedom to operate independently whenever the situation calls for it.

The drone array, which is a controlled maneuver swarm with aerial combat drones as elements, is an example of a true robotic military platform of the second wave. It embodies and delivers all six of the inherent tactical advantages of robotics. It can provide a commander with virtual presence and the ability to conduct operations anywhere within a vast sphere of influence extending many kilometers in radius. It capitalizes on the power of precision weapons to deliver massive asymmetric combat power anywhere at choice. It moves swiftly and elusively across the battlespace regardless of terrain while presenting a bare minimum of target. It can conduct the details of engagements at superhuman speed enabled by AI. It cannot be destroyed by a single blow of nearly any weapon, and if hit, can gracefully shed any damaged elements and continue its mission with only gradual loss of function and no loss of human life. Along with its unmatched mobility, it can land to conserve power and remain in position for days if the mission requires. With all these advantages, the most effective platforms for future ground combat may not be ground vehicles at all.

THE RETURN OF MANEUVER ON THE PRECISION BATTLEFIELD

We have identified the drone array as an iconic example of a robotic platform exploiting the new robotic combat regime of the atmospheric littoral. Now we can get more specific about how it can restore maneuver on the precision battlefield and outmatch enemy forces in future combat.

Drone arrays are built to survive on the precision battlefield. The drones, perhaps two hundred times less massive than a tank, are much harder to find and fix. Their speed makes them largely immune to precision artillery and other indirect fires. This elusiveness is amplified by their ability to move under treetops, behind buildings, and down side streets out of line of sight, which lets them evade the latest short-range air defenses. Unlike ground vehicles, they are immune to anti-tank ditches, rivers, minefields, and even the infamous mud that sometimes brought ground movement to a halt in Russia and Ukraine. Unlike expendable robotic munitions, they can execute behaviors such as flanking maneuvers, breakthroughs, and envelopments, restoring the tactics that make offensive maneuver so decisive. They go far beyond the limitations of those classic tactics by extending the art of maneuver into the vertical dimension. They work in coordination with robotic precision weapons such as loitering munitions, providing a synergistic combination of fires and maneuver. Because the atmospheric littoral liberates movement from the constraints of terrain, it provides a ground commander an entire new regime of tactical options that were never possible before.

Imagine a single array of perhaps a dozen combat drones attached to an existing small unit, such as a US Army or US Marine infantry company. A single human soldier or marine controls the array in close coordination with the unit commander. A display screen, or a set of VR goggles, allows the controller to see the video from any drone's cameras. The controller can switch the view from drone to drone, or from visible light to other cameras such as infrared or night vision.

Let's follow this infantry company as it advances through a campaign, using the drone array to achieve advantage against today's high-end adversaries in almost every phase of maneuver.

Movement to Contact

During this type of maneuver, the unit is moving and looking out for the enemy. As the infantry company moves toward a village, for example, the drone array advances in a widely spaced skirmish line a kilometer or two ahead of the camouflaged human troops, screening everything with their sensors. Ubiquitous tiny ISR drones watch the landscape from high above, while the combat drones skim ten meters off the ground, just higher than the houses. Individual drones use AI to search their live video feeds for hostile targets near the buildings ahead while the controller toggles from drone to drone, examining the village from different viewing angles. A drone's AI sends automatic notice of a threat identified as a shoulder-fired anti-aircraft missile as the missile races toward it from an open window. The drone's automatic evasion routine causes it to zip downward momentarily to nearly touch the ground, and the missile impacts the earth behind the drone. Alerted to the enemy presence in the village, the commander's human troops remain unseen while the drone array provides many options. The commander could withdraw without danger, or if he or she chooses to attack, the array could sweep into the village or strike it from a distance.

Shaping Engagements

When the unit encounters an enemy force, quick initial moves can give one side or the other the advantage for the entire engagement. The close-up, ground-level view from the drones reveals a powerful enemy mechanized force waiting in ambush within the village, armed with early twenty-first-century equipment. A precision-guided missile from a drone flies through the previously identified window and destroys the fighting position inside, and the combat drones fan out to take advantageous positions around the village. As the enemy realizes it is facing a drone array, it unleashes the precision-guided artillery waiting several kilometers behind, but the shells cannot target the elusive drones and fall like sledgehammer blows missing a swarm of bees. Instead, the drones have the enemy forces in a crossfire, and their missiles swiftly hit several armored vehicles hiding under cover. Their precise automatic weapons fire forces the enemy's sentries back from their positions in upper-story windows. Despite the enemy's superior strength, the friendly unit has seized the initiative while denying the enemy the ability to hit back. The friendly unit commander has the advantage in situational awareness and tactical position.

Attack: Combined Arms Assault

Effective assaults seek to combine the action of different classes of weapons for synergistic effect. The commander decides to press the attack. First, several loitering munitions fix and destroy the enemy artillery vehicles beyond the village. Then the infantry launches dozens of small signature-seeking munitions from its compact VLS cells. From their low-angle forward positions, the combat drones automatically cue the approaching munitions with target locations and target types for vehicles hiding inside structures, under trees, and with other visible enemy presence. As the munitions arrive, the array concentrates and sweeps in from an unexpected direction. Benefiting from their cues, munitions quickly destroy some of the enemy unit and paralyze it under fire. Meanwhile, the array enters the village and isolates position after position with concentrated converging fire from missiles and automatic rifles. As enemy elements try to retreat from the array, they expose themselves to the hunting munitions, which all find enemy targets within a few minutes. The survivors of the enemy force flee on foot, their mechanized ambush having never caught sight of the friendly ground elements. Using the drone array, the commander scours the village for pockets of resistance and then allows the array to recover to the rear, where it automatically refuels and rearms. When the friendly infantry troops enter the village later, they capture dozens of unused antitank missiles and MANPADs.

Attack: Penetration and Exploitation

After evicting the enemy force from the village, the unit prepares to assault a strongly fortified and seemingly endless defensive line protecting the territory the enemy has illegally seized. Frontal attacks against prepared defenses are some of the most difficult tactical challenges. Such operations attempt to break through at a point and then expand the breach and threaten the enemy's rear areas. After providing a detailed ISR picture, the drones converge from many directions at treetop level. Flying over the enemy's minefields, they give the enemy ground forces and their air defenses no time to respond. Ignoring the trenches and obstacles, they sweep overhead. Considering the military setting, the operator gives them increased targeting autonomy, and they strike air defense and armored vehicles, command trailers and communications equipment, and other targets under high-speed AI control. Their speed and low altitude make them nearly impossible to target from the ground. With the enemy's prepared ground defenses worthless and outflanked, many of its troops attempt to escape their positions. Those that refuse to flee the trenches are outgunned as drones fire down the length of the trench lines. The unit uses the drone array to roll up the enemy defenses and expand the breach by attacking its exposed flanks.

Attack: Flanking and Envelopment

Effective maneuver provides the ability to surprise and disrupt an opposing force by approaching from an unexpected, seemingly inaccessible direction. Days after advancing into enemy-held territory, the unit finds itself fighting in an urban neighborhood surrounded by multistory buildings. Friendly infantry teams on streets just one block apart are unable to see each other or support each other with fire. Enemy troops fortified in upper floors seek to fire down from commanding positions. Other enemies advance down side streets, looking to surprise the infantry. The commander sends the drone array upward to the level of the rooftops and directs precision fire that clears enemies from the upper floors, negating their positional advantage. Next, the array crosses over the top of the intervening buildings and ambushes the enemy forces on the adjacent street, enveloping them from above. The drones transmit precise intelligence from their cameras and AI target-recognition systems to the friendly troops below. Faced with precision fire from the drones performing a vertical envelopment and the infantry moving to surround them on the ground, the enemy forces are forced to withdraw with heavy losses.

Infiltration and Interdiction

Friendly forces often need to strike enemy forces or supplies that are moving well behind the front line. As the infantry clears the city block and resumes its advance, its drone array autonomously refuels and rearms at a logistics point a few blocks to the rear and is quickly back on station. Then, the unit receives an ISR warning that the enemy is preparing to launch many loitering munitions from a city square a kilometer away. With no time to coordinate air support, the ground commander dispatches the drone array directly. Completely immune to the complex urban terrain, the array arrives at the square in less than ninety seconds and catches the launchers and enemy troops by surprise. A few moments of coordinated, AI-targeted precision missile fire leaves all the launchers in flames and the enemy troops scattered. Noting that the enemy's short-range air defenses are responding, the array controller orders the array to split up and infiltrate back to the unit individually between the buildings. Like water, they filter back through the gaps and fissures of the city and then reestablish their formation.

Defense: Area Defense

Defending a large area against enemy attack can stretch a small force. After helping to win the urban battle, the infantry unit deploys to another part of the theater where it is assigned the difficult task of holding a valley ringed by hilltops against enemy insurgent attack. Enemy units try to assault and overrun small outposts that are located on the hilltops where they cannot be easily reinforced. However, the commander uses the drone array as a mobile quick reaction force. It flies across the valley to reach any hilltop in minutes and delivers instant effective mass to seize the initiative and repel the attacking forces. During some periods the commander has the array land on a distant hilltop to watch for enemy activity and prevent the enemy from stealthily occupying the position. Because the drone array lets the unit cover many locations and strongly defend any hilltop within minutes, the enemy is forced to abandon its strategy for capturing the valley.

Defense: Mobile Defense and Retrograde

Even the most effective force may sometimes choose to retreat, and the drone array provides great advantages then as well. The infantry unit faced its worst situation of the campaign when it was threatened with encirclement by a force ten times its size. Faced with this scenario, the unit places the drone array between itself and the advancing enemy. Ordered to swarm by force, the array uses rapid advances and withdrawals, unpredictable movement, and precision fire to hold the enemy at bay while the manned forces withdraw safely. Unlike a force on the ground, the drone array cannot become pinned down by enemy fire and can always disengage at will. However, to delay the enemy's offensive, the commander orders the array to hold its position and fight. Over the next several hours the array slowly depletes its logistics point while gradually losing drones one by one to enemy fire. The infantry unit moves far away while the drone array inflicts increasing losses on the frustrated enemy. By the time the enemy's battered forces destroy the last drone, the robotic "suicide mission" has severely degraded the enemy's offensive. The friendly infantry unit took no casualties, and replacement drones arrive soon to reconstitute the array.

These scenarios are all simplified. In real combat, an infantry unit like this might benefit from support from long-range artillery, missiles, and other precision munitions; friendly airpower; and ISR from space and other assets. It might have support from neighboring units instead of constantly fighting on its own. However, the scenarios illustrate the power of the drone array and the versatility of maneuver in the atmospheric littoral. The drone array gives the unit the benefits of most of those supporting capabilities and more, from within its own resources. It provides overmatch in almost every tactical situation. In addition, in the scenarios, the enemy forces rarely encountered the infantry unit's human soldiers or manned vehicles. It executed the most demanding maneuvers while the soldiers remained out of contact and largely out of danger. In effect, the drone array lets the unit conduct close combat at long range.

In these scenarios, the drone array was attached to an infantry company, but drone arrays could be attached to other kinds of units. If attached to armored units, they could realize and extend the advantages of manned-unmanned teaming that were glimpsed in the Battle of Kursk decades ago. That could extend the viability of today's armored formations. If attached to navy or marine amphibious units, they could easily cross water barriers to project power ashore from ships and deliver powerful maneuver during combat in island chains and archipelagos.

The atmospheric littoral is a new tactical regime. Drone arrays provide capabilities that are fundamentally unobtainable through existing means such as ground vehicles, manned aircraft, or fixed-wing drones. Their mobility, lethality, flexibility, and attritability will make them the go-to option to accomplish many maneuvers. This will be the case in both lethal and nonlethal operations. As a result, we can anticipate that drone arrays may be favored over other kinds of platforms and may take over more and more roles. Before long, they may displace traditional platforms and dominate maneuver combat in the robotic age.

THREE AREAS OF CHALLENGE

What prevents us from fielding drone arrays right now? The technology to build armed drones themselves is here, and a purpose-built combat drone could be designed and built tomorrow that is greatly superior to the improvised civilian-drone-based models common today. Indeed, individual "flying remote weapon station" drones could quickly add a great deal of new capability to existing units. However, before the drone array can take its place as a revolutionary maneuver platform of the robotic age, challenges need to be overcome in three main areas: command and control, autonomous logistics, and combat AI. These address the most important lessons from the history of robotic warfare.

Command and Control

Today, operators can control individual drones, and pre-programmed automated systems can control large numbers of drones such as during scripted light shows. However, controlling many drones at once, in real time, is burdensome. As we know, burden has been the Achilles' heel of robotic platforms for a century. In addition, the complex behaviors expected of a drone array simply cannot be performed using individual control. We must use new means that can allow individual controllers to direct entire drone arrays quickly and intuitively in a changing and unpredictable environment.

That kind of intuitive control requires a high degree of situational awareness. An operator needs to be able to see where the array is within the battlespace and also be able to see the environment from the viewpoint of individual drones as needed. The operator's interface must be immersive and easy to use. The good news is that a generation of video gamers are already comfortable with these kinds of interfaces. A whole genre of real-time strategy and battle-arena games such as the *Warcraft*, *Starcraft*, *Age of Empires*, and *Total War* series allow individual players to control groups of forces, like companies of soldiers or squadrons of airplanes, with single clicks on a real-time battle map. The computer handles the detailed behavior of the individual elements within those groups automatically. Similarly, players of first-person shooter games are comfortable seeing the world from the vantage point of a virtual player embedded in a 3-D landscape. The intuitive control of drone arrays and other dissociated robotic units will involve implementing those familiar control interfaces in the real world.

Efficient control also depends on the ability of drones to assume and maintain formations, move as a group, avoid obstacles, and locally coordinate without explicit external control. The array behaviors that researchers have demonstrated so far must be extended to more complex military-specific maneuvers in open landscapes and in complex urban environments. This work is benefiting from commercial investment in applications such as drone package delivery.

Command and control also depend on communication links. A human drone controller can provide direction from a safe location in a concealed vehicle, an underground bunker, or even from a continent away. However, the sensor data needs to be able to reach the controller, and the controller's commands need to reach the drones. Jamming drone communications is not always simple in practice, as pointed out earlier, but jamming capabilities are improving. However, new technologies allow for tremendous levels of redundancy and security. Consider that a late-generation smartphone contains digital radios operating in the 3G, 4G, and 5G bands, plus the 802.11 Wi-Fi band, plus Bluetooth. The phone also contains a GPS receiver, plus a miniature inertial measurement unit for sensing the phone's orientation and movement. All that costs only a fraction of the phone's price of $1,000 or less. That hints at how many independent communication and control mechanisms could be embedded into a combat drone.

Such capabilities, and increasing levels of autonomy, will make future drones much more difficult to jam. In addition, many military radio communications are concerned with jam resistance and security over distances of fifty or one hundred kilometers or more. At the short ranges needed for tactical use in the atmospheric littoral, often a few hundred meters or less, drones can maintain almost un-jammable, self-healing mesh networks using directional radio or laser communications. There are plenty of options for secure communications, but they will need to be customized and ruggedized for use in combat drone arrays. In addition, the designers of datalinks have the advantage of being a move ahead of the adversaries who have to come up with ways to disrupt those new datalinks. But the enemy will adapt. Friendly forces must be prepared to regularly update their datalinks to overcome the enemy's attempts to disrupt them.

Autonomous Logistics

The greatest burden of operating robotic systems often comes when they need to be recovered and prepared for a new mission. Drones will eventually run short of fuel, whether liquid or electric, and be depleted of ammunition. Currently, refueling and rearming unmanned systems can consume an entire unit's attention between missions, which might be impossible during active combat.

Therefore, drone arrays must be able to refuel and rearm without human intervention. Just as with soldiers and manned vehicles, they must be provided with replenishment locations, but otherwise, they will refuel and rearm themselves. We expect this from domestic floor-sweeping robots that return to their charging stations when their batteries run low. Military engineers have developed a similar capability for small drones.[180] Powerful combat drones should do much more. Host units can provide reservoirs of fuel and ammunition, either a few blocks to the rear, or placed by drones in relatively inaccessible areas such as the roofs of buildings. With the help of some clever and relatively simple design, individual drones can dock with those resupply points.

For instance, a pressurized fuel bladder with a docking port on top can allow a hovering drone to insert a refueling probe and refuel in seconds, like a hummingbird sipping nectar from a flower. A frame with full weapon magazines and missile tubes could allow a drone to drop its depleted ones and then snap new ones into place by landing on top for a moment. At a modest speed of thirty miles per hour, in one minute a drone can travel the distance of seven large city blocks to the rear to refuel and rearm at a logistics point. In another minute it can be back in the fight. With this capability, the endurance of atmospheric littoral drones can be practically unlimited, as with today's combat aircraft that are provided with air-to-air refueling.

Drones will sustain combat damage. Self-repair is likely to be impractical, but drones can be built to tolerate damage and stay in the fight despite losing parts to enemy fire. For instance, engineers have demonstrated technology that enables multi-rotor drones to adjust autonomously to the loss of a rotor.[181] With damage-tolerant design and some AI, individual drones can compensate for damage and fight on without need of human intervention during a battle.

Combat AI

This third challenge area recognizes that the combat environment and combat missions demand different kinds of AI than is needed for common civilian applications like natural language processing or self-driving road cars. To minimize burden, arrays will need to accept and execute commands that are similar to those given to manned units and interpret them in a context-appropriate way. That may include commands like "move here" or "destroy this target." Because they will cooperate closely with human ground forces, they may need to receive local commands from the ground using laser target designators or even hand signals.

Beyond the needs of AI for navigation and control, the context of combat in the atmospheric littoral defines certain AI problems that should drive future development. Even though tough issues of judgment and decision-making can be handled with the help of a human controller, the drones will need to be able to address difficult AI problems on their own. Some examples of specific AI challenges include:

- **Target acceptance.** Automatic target recognition is the ability of a sensor system to recognize and flag potential targets based on predefined characteristics. Drone arrays will need a related but distinct ability to accept the designation of a target by the operator and understand that target's boundaries and what it consists of: a static object, part of a static object (for example, a window of a building), a road, a vehicle on that road, an area of ground, and so on. Human soldiers often understand the target based on context, and AI will need to develop a similar capability.
- **Target keeping.** Drone arrays must retain the designation of a target even if it changes orientation or appearance, gets obscured by smoke or weather, or disappears behind another object. The ability to understand, for instance, that a tank is still present behind a

building even though it is not visible anymore is called "object permanence," and it's a key step in the cognitive development of young children.

- **Damage assessment.** After attacking, drones need the ability to determine whether their target has been destroyed so that they do not continue to needlessly attack an already neutralized target—or stop before the target has been neutralized. Damage assessment currently relies on human judgment, but drones will have to do it for themselves.
- **Incoming fire awareness.** One reason why robotic vehicles were historically vulnerable on the battlefield is that they did not know when they were under attack. To survive, drone arrays will need to be able to detect when they are under attack or when a drone has been hit and take appropriate defensive action, such as evasive movement, while alerting the controller.

Because the air will likely be full of enemy drones, combat AI will ultimately need to incorporate drone vs. drone combat, so those enemy drones can be destroyed whenever they are encountered. Inevitably, adversaries will field their own drone arrays. So, control of the new key terrain of the atmospheric littoral will be contested between competing drone arrays, with contributions from ground-based short-range air defenses. This may be simpler than ground combat in some ways, because the air environment is simpler. However, the speed and complexity of multi-drone air-to-air combat will require that it be conducted mostly by AI. Researchers have anticipated this and have demonstrated swarm vs. swarm drone "dogfights" using simulated weapons starting in 2017.[182]

Relatively straightforward tactical problems for AI will push the state of the art for some time, because what is easy enough to demonstrate in a computer simulation or a laboratory can be devilishly hard to do in the real world. This is especially true in combat, the most challenging of all real-world contexts, where adversaries are doing their utmost to confuse, obscure, and interfere, and weather and other factors can make even the simplest things difficult.

In 2024 the Army Futures Command hosted an industry symposium on the Air-Ground Littoral in a hip office complex in Austin, Texas. The energy there was palpable, with an overflow audience ranging from top Army generals to defense industry executives to independent innovators just returned from the front lines in Ukraine. Five years after I introduced the atmospheric littoral concept in 2019, it was inspiring to witness the eagerness to address this great strategic opportunity. There is so much to be done.

And not much time to do it. The challenges and opportunities of drone arrays and the atmospheric littoral should be worked out via field experiments and rapid iteration. Individual combat drones can add useful capability today, so militaries should move quickly. Military services should procure small numbers of purpose-built militarized drones for experimentation and use the results of the experiments to refine requirements for the next generation of drones. Experiments should bring together the latest prototype hardware and software, new doctrinal concepts, and forward-thinking warfighters. There is no time to lose, as foreign militaries and militant groups are already developing drone-based robotic warfare techniques as quickly as they can, under the pressure of ongoing conflicts.

HOUR TWO OF THE SIX-HOUR WAR: THE FARTHER FUTURE

The sergeant peers out from his hiding place at the edge of the pine forest. The sudden outbreak of war has caught him there, and now he can't move. Somehow, he's survived this long, and he knows that keeping under cover is his best bet for staying alive. However, he has never witnessed a robotic field battle with his own eyes before, only on video screens. His curiosity is unbearable. He can see smoke rising from the nearby wreckage of a cluster of American security vehicles hit during the initial battle, a while earlier. Holes had been punched cleanly through each of them in similar places: surgical hits from signature-seeking loitering munitions. Farther away, other pillars of smoke rise, the aftermath of the strikes by waves of containerized precision weapons from both sides that swept the landscape of every visible target. The hazy sky is streaked with peculiar contrails, evidence of an air battle high overhead.

He crawls a meter forward for a view across open land that slants away for a few kilometers toward other copses of pine trees. He's reluctant to move from under the foliage, imagining the unseen electronic eyes peering down from above. Though he can't see anything moving, the sergeant knows the battle is evolving quickly.

He watches as a missile from the enemy's direction reaches the area overhead and separates into more than a dozen small drones. The fact the missile can reach so far shows that the friendly air defense is in bad shape. As the drones hover down into the tree line a few hundred meters away, the sergeant covers his head and tries hard to remain motionless. For the next couple of minutes, sharp bangs echo as each drone finds something that matches its target parameters. The explosions come dangerously close to him before petering out. He exhales with relief and looks up.

Something overhead catches his eye, a strange aircraft flying swiftly over the landscape. He can't discern its shape, but it seems to glitter in the pale sun. Then with a flash, a missile streaks upward from one of the clumps of trees. In a moment it reaches the aircraft—but the aircraft scatters into many tiny pieces like a school of fish, and the missile passes through. The pieces swerve sharply in concert and then rejoin, racing lower in a new direction that takes them almost above the sergeant. As they pass with a shrill whine, he sees, past the pine boughs overhead, the cluster of silvery shapes, like a handful of daggers flying so close together that they nearly touch. Something new from the Air Force, he supposes.

A line of combat drones rises from the ground far ahead, changing color from the buff of the grassland to the pale blue of the sky. A streak of light shows a missile leaping from one drone, and it's quickly joined by two others from its neighbors. Almost faster than he can see, the three converge on the same target, overwhelming the active defenses of the enemy air defense vehicle that had fired from somewhere behind the distant tree line. A cloud of fire and smoke rises, and the boom comes a few seconds later.

Then the sergeant notices the dreadful hum of many drone motors behind him. Adrenaline surges, and he braces for the impact of bullets cutting through his flesh. That doesn't come, though surely the drones have already seen him. He presses his body to the ground as an entire array cruises overhead between the tree trunks, bristling with weapons. They are so close that he can see the pattern of chromatic tiles on their surfaces.

"Don't move, Sergeant," a female voice commands from the nearest drone as it passes. He recognizes it as Lieutenant Jones, the drone controller for the battalion's third company. "Stay under cover and we'll pick you up when the command section moves forward." The drones move into the open, then race away, shifting color and opening their formation as they go.

He backs slowly toward his hiding place in the forest. Across the landscape, several of the battalion's other drone arrays sweep forward as the army goes on the offensive. Here and there a drone corkscrews into the ground, hit by some kind of enemy weapon, but the arrays shrug off the losses and race on. A new wave of loitering munitions appears overhead, moving in the same direction. Beyond the distant trees, explosions erupt like strobe lights. He puts his head down amid the pine needles and waits. This battlefield is no place for a human.

6

DANGEROUS THINKING

Blind Spots, Mistrust, and the Future of Combat AI

During 2019 and 2020, two parallel competitions highlighted both the promise and the challenges of combat AI. Following the remarkable news that the powerful deep-learning AI system AlphaGo defeated the human grand masters of the board game Go, the Defense Advanced Research Projects Agency, or DARPA, conducted the AlphaDogfight Trials. DARPA sought to apply the same deep-learning techniques that had allowed AlphaGo to beat human champions in a one-on-one board game to another kind of complex game: a one-on-one aerial dogfight between fighter planes. Eight teams from industry and academia developed AI systems that applied the deep-learning approach within a simulated, virtual air combat environment. Like AlphaGo, they ran millions of virtual contests within a digital world, allowing the powerful system to learn what worked and what didn't.

The teams faced off against one another in virtual one-on-one dogfights, each flying a simulated F-16 fighter. The winning team, from Heron

AI, then faced a real, experienced US Air Force pilot flying a virtual F-16 using a simulator. The AI beat the human pilot five wins to zero.[183] Just as with AlphaGo, the AI found unusual and unexpected tactics, such as flying directly at the human-piloted jet as in a game of "chicken," and blasting it with cannon fire at point-blank range just before collision. The DARPA program manager cautiously announced that the AlphaDogfight Trials had demonstrated "that an AI agent can quickly and effectively learn basic fighter maneuvers and successfully employ them in a simulated dogfight."[184] Encouraged by those results, the Air Force accelerated programs, such as Skyborg and Project VENOM, to develop and test AI codes capable of controlling real combat aircraft such as F-16s and forthcoming jet-powered drone wingmen.[185]

At almost the same time, the worldwide Drone Racing League launched its AI racing division. Most Drone Racing League races feature human pilots wearing VR goggles steering first-person-view racing drones through real-world courses at dizzying speeds of up to 120 mph (190 km/h). The AI Robotic Racing Circuit consisted of four races in its first season. Defense company Lockheed Martin sponsored the season with a million-dollar grand prize plus the chance for the winner to go head-to-head with the racing league's human champion for another quarter-million-dollar bonus.[186] Each team from industry and academia brought an autonomous racing drone with onboard cameras and sensors, backed by powerful machine vision and navigation systems.

Like AlphaGo and the AlphaDogfight teams, each so-called AlphaPilot team trained its AI system in countless virtual races until it achieved amazing performance. The difference was that, in the circuit, the AI drones would have to navigate a simple but unfamiliar course in the physical world. The league's spokesperson said that in contrast to putting human players into a virtual world, the event would bring "virtual players into the human world . . . taking a computer player and putting it into a human sport, and seeing if it can beat human players on a level playing field, with the same hardware and the same courses."[187]

As in human-piloted drone races, the drones had to pass through large square gates positioned along the course. Where the human races featured groups of drones racing together in complicated courses full of obstacles, neon lights, and blind turns, the first AI course, at the Drone Racing

League competition in Orlando, Florida, was simple, with just a few large gates in an otherwise empty arena. Nonetheless the crowd was filled with curiosity and suspense. Pulsing music shook the air. Lockheed Martin logos festooned the arena, and real-life military drone pilots were there in uniform, surrounded by excited kids.

The competition began. One at a time, the drones took off for a timed run of the track. Each rose into the air and peered at the course ahead, then moved slowly forward.

One by one, they wobbled and crashed.

A single drone managed to pass through the first gate, which was only ten meters or so from the launch point, before blundering into the second gate just beyond and falling to the floor. The second event, in Washington, DC, produced a similar result. By the third event, in Baltimore, one drone was able to make it through two gates. In advance of the final in Austin, Texas, the teams were allowed to conduct more thorough and specific deep-learning training runs in a realistic physical test course. The winning team from the Netherlands was able to pass through all four gates in nearly a straight line and then fly into a safety net, reaching a top speed of 20 mph (33 km/h).[188] The final race against the top human champion was hardly a contest at all, and the Drone Racing League canceled future seasons. As the Dutch winners concluded, the technology was able to beat humans in lab simulations, but "there is still a huge challenge to beat humans in more complex and unreliable environments."[189]

In 2023, a team from the University of Zurich and Intel Corporation revisited the challenge. They built a very similar circuit of seven gates in an empty arena. Their AI racing drone, called "Swift," was able to complete full laps of the circuit and even outrace top human champions. However, the team had adjusted the problem to be more amenable to deep-learning methods. They made the real world look as much like an idealized virtual one as possible. They precisely modeled the course and its gates in a digital simulation that allowed the AI vast numbers of finely tuned training runs in advance. The AI system still struggled when something subtle changed from the digital simulation, such as a change in a gate's position, or even a change in lighting.[190]

The previous two chapters focused on how military robotics will transform the physical character of combat systems. However, future advances in military robotics depend on what is not visible: advances in AI. The previous chapter highlighted some examples of specific combat AI challenges dealing with navigation, targeting, situational awareness, and autonomous logistics. As the robotic revolution proceeds, the number and variety of practical challenges is multiplying.

By the time you read this, any details about the state of the art of AI at the time of this writing will be obsolete. Therefore we will focus attention on the forest and not the trees—the important challenges regarding military AI that are larger issues than the latest AI demo or technology release can address. The central problem is that current AI technology is powerful in many ways, but it also has fundamental weaknesses. Those weaknesses are dangerously aligned with the difficulties of the real-world combat environment. Nonetheless, powerful human psychological biases and blind spots encourage us to ignore those weaknesses, opening the door to mistakes and potential disasters.

THE REALITY OF MILITARY AI MISTAKES

In military operations, bad things happen when a robotic system misinterprets something in the real world. Active defense systems have provided many cases in point. For the past decades they have been the most autonomous of armed robotic systems, because their purpose is to detect and defeat incoming threats like homing missiles on short notice. Like a car's robotic airbag system or emergency braking system, they often must detect a dangerous situation quickly and respond automatically.

For example, during Operation Desert Storm in 1991, the battleship USS *Missouri* was sailing in the Persian Gulf with several escort ships, striking Iraqi targets with Tomahawk cruise missiles and sixteen-inch gun shells targeted with the help of Pioneer drones. Iraqi forces launched two old anti-ship missiles, upgraded versions of the Styx missile from the 1960s, at the *Missouri* from launchers on shore. One of the missiles malfunctioned, and the other was shot down by a Sea Dart air defense missile from one of the escorts, the British frigate HMS *Gloucester*. However, when false reports arrived of other missiles approaching, the *Missouri* fired clouds of radar-reflecting chaff to confuse the missiles' homing radars. One of the escorts, the American frigate USS *Jarrett*, had set its Phalanx anti-missile Gatling gun system in auto-engage mode. The Phalanx falsely targeted the clouds of chaff between it and the battleship and loosed a stream of bullets that strafed the *Missouri* before a quick-thinking crew member shut the Phalanx system off.[191]

During the invasion of Iraq in 2003, similar misjudgments caused Patriot air defense missile batteries to accidentally shoot down a US F-18 fighter and a British Tornado strike jet in separate incidents.[192] Another such mishap occurred when a US Navy ship shot down an F-18 in December 2024.[193] In some such cases, the defensive weapon wasn't in full autonomous mode, but because the information processing was done automatically, the human crew accepted its incorrect judgment and authorized the system to fire. In a chilling example from the Soviet Union, an automated satellite-based strategic missile warning system misinterpreted the sunrise reflecting from clouds as a wave of US ICBM launches. It advised an immediate massive nuclear counterattack. Only the quick intervention of a single Soviet officer prevented the recommended launch, an incident that was dramatized in a 2013 documentary entitled *The Man Who Saved the World*.

Those examples involve information processing systems that predate modern machine learning–based AI. However, examples abound of supposedly sophisticated modern AI target recognition systems failing in tests. For example, a machine-vision system trained to recognize approaching soldiers was easily evaded by marines who held tree branches in front of themselves or wore cardboard boxes, simple tricks that confused the highly trained AI.[194] Other target recognition systems that leveraged the power of large language models like ChatGPT were spoofed when testers wrote the name of a false object across the target, like writing "school bus" across the top of a tank. That simple trick exploited the language recognition abilities of the latest AI as a weakness.[195]

More and more military applications are indeed coming within the power of AI. However, many of the challenges with AI in a combat environment are not simply technical hurdles that can be overcome in the next release. Instead, they are related to fundamental weaknesses with current AI methods and their application to combat.

CURRENT AI HAS FUNDAMENTAL WEAKNESSES

The current wave of AI excitement is driven by a particular approach that has come to dominate AI research and development. That is AI based on machine learning, and to be more specific, the machine learning approach known as "deep learning." Machine learning was one of several competing approaches to AI for many years, and deep learning is a particular subtype of machine learning that rose to preeminence around 2010.

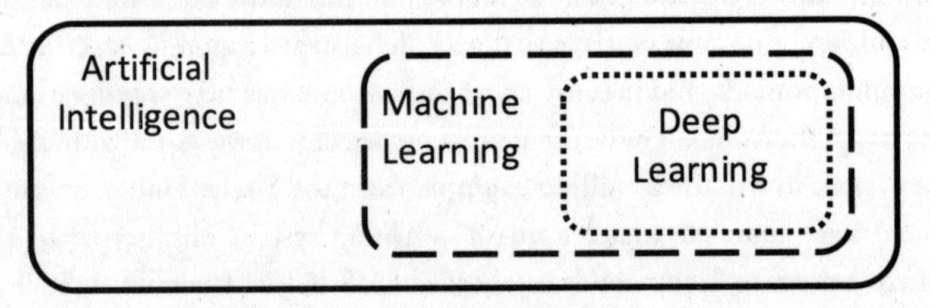

Deep learning is one approach to machine learning, which is one approach to artificial intelligence.

Like robotics, AI was born with warfare and defense applications in mind. Researchers at the University of Pennsylvania built the first large-scale digital computer in the US, called ENIAC, in 1945 to perform weapon guidance calculations and help in the design of hydrogen bombs. Industrial competitors quickly followed, including the UNIVAC 1101, which the military funded for breaking Soviet military codes, and the IBM 701, the company's first mainframe, which IBM marketed as the "Defense Calculator." A lot of early AI work was funded by the Department of Defense's Advanced Research Projects Agency upon its founding in 1957.

Much of this early work used symbolic logic in a direct attempt to build logical thinking machines. However, some computer scientists thought that building digital brains from the bottom up was the way to go. They started with single electronic neurons that could react to simple impulses and then connected them to form networks, much as the human brain evolved from the most primitive organisms' nervous systems. The Office of Naval Research funded the first working examples of simple neural networks. They were the so-called "perceptrons" invented by researcher Frank Rosenblatt around 1960.

Using digital neural networks, scientists could train computers to perceive and recognize stimuli and patterns, much like a bee recognizes the shape of a flower. Instead of coding an algorithm directly using formal logic, the neural network learned through repeated exposure what weighting factors to apply to each neuron in its network to best identify a particular stimulus or pattern. By stacking digital neurons into layers, they could make more sophisticated interconnections and provide features such as "backpropagation" that allowed for better performance. While the machine learning techniques did enable impressive feats of simple pattern recognition, it became obvious that even a machine learning system with a bee's brainpower would require computers with neural networks thousands or even millions of times larger than they could feasibly build with 1960s electronics. The machine learning approach stagnated. Formal coding produced vastly more useful software for decades.

However, over time, formal symbolic approaches to building artificial intelligence ran into roadblocks. Humans seemed to have an almost infinite and hard-to-define set of understandings and concepts about reality that underpin our judgment and common sense. Symbolic approaches produced software that was excellent within very specific areas of logic, such as performing specialized medical diagnoses or guiding missiles, but could not be generalized beyond that. Several "AI winters" ensued as hopes for various formal methods and machine learning methods faded.

Meanwhile, engineers invented the semiconductor microchip. Microchips followed Moore's Law, packing exponentially greater numbers of electronic devices onto each square millimeter of silicon over time and driving the cost of computing down and down. Eventually, the previously unfeasible happened. Around 2010, computers millions of times more powerful than those of the 1960s became inexpensive enough for real machine learning applications.

In 2012, a graduate student named Alex Krizhevsky won a computer image-recognition contest using an AI system based on machine learning.[196] His system was a small but powerful computing engine that incorporated commercial graphics processing unit, or GPU, chips. They provided millions of simple devices that he repurposed as digital neurons. He trained the system using a deep archive of sample images that had been labeled in advance. Using this deep set of labeled training data, his system could basically train itself. The exact weighting factors it chose for the myriads of digital neurons in their many layers was impossible to know, but with additional training it got better and better. His system won the contest by a wide margin.

Within a few years, researchers abandoned dozens of other approaches to image recognition in favor of deep learning. They quickly applied the technique to more and more problems, leading to the current renaissance in AI. A great many problems, it turned out, could be structured as pattern-recognition problems. They could be solved using simple methods combined with masses of computational power and enough solved examples. With sufficient computing power and good training data, the mysterious process of self-training produced amazing results in area after area. The approach was brute force: If a system didn't produce the required performance, researchers built a bigger system and tried again. Like the race among physicists to build bigger and more powerful particle colliders, every leap in the size of deep-learning systems seemed to reveal more amazing new results and behavior, even if the understanding of what was going on often lagged far behind.

The remarkable power of deep learning comes with fundamental weaknesses. Researchers summarize the most pervasive of them as brittleness, opacity, and greed.

Brittleness means that the apparent intelligence of a system is strictly bounded by the data on which it was trained. Because its performance is not based on underlying logical concepts, it can fail suddenly and dramatically the moment the situation exceeds its training set. For example, some tests of drones with navigation systems based on deep learning have seen them maneuver smoothly until they sensed some unexpected factor, upon which they swerved out of control without warning.

Opacity means that the behavior of deep-learning AI systems is hard to predict, or even to understand after the fact. Their performance is sometimes better than expected, a happy accident. However, when they fail it can be hard to learn why. Sometimes the only option is to retrain them using better data and hope for a better outcome.

Greed means that deep-learning systems require huge amounts of computing power and data. Many run on giant data centers the size of industrial factories that contain tens of thousands of GPU cores, each of which costs tens of thousands of dollars. Each such data center can consume as much electrical power as a small city.[197] Accessing that computing power from the field requires an assured network connection to cloud services, something difficult to achieve in the heat of battle.

In addition, many applications simply can't furnish the massive sets of labeled training data that deep-learning AI models need. The largest and most commercially important AI services, generative AI tools such as large language models and AI image generators, use the kinds of training data available everywhere across the internet, such as digital text and images. ChatGPT, which is short for "chat generative pre-trained transformer," was released to global attention in 2022. It was trained on three hundred billion words of internet data.[198] More recent models consume many times more data. Vast datasets like that are not going to be available for most specialized military problems.

WAR IS THE MOST AI-UNFRIENDLY ACTIVITY

The fundamental weaknesses of deep-learning-based AI are so inauspiciously aligned with the fundamental difficulties of war that warfare may be the last domain in which AI can successfully operate unsupervised. Deep-learning-based AI thrives on problems that can be clearly bounded, so that its training data can cover all potential factors. Its brittleness causes failures when it faces unexpected situations that its training data did not encompass. Its greed requires large sets of specific, well-formulated, and curated training data, and assured access to digital resources. In short, deep learning works best in orderly, easily digitized, and carefully bounded virtual environments.

In contrast to what deep-learning AI requires, war is perhaps the most volatile, uncertain, complex, and ambiguous of all environments. It takes place in a stubbornly analog physical reality. The classic scholars of war such as Carl von Clausewitz expounded on war's intractable fog and friction. *Fog* is the miasma of information scarcity, overload, and unreliability that confuses everything in war. *Friction* is the tangle of unexpected and confounding factors that make even the simplest things difficult.[199] Unpredictability is unavoidable, and it is said that no plan survives first contact with the enemy, who will do everything in his or her power to create more surprises and chaos. Moreover, much of the fog and friction is due to what are often called "unknown unknowns," things that we don't imagine could be an issue and cannot be anticipated.

So war is characterized by the very qualities that are hardest for deep-learning systems to manage. In some circumstances, carefully applied AI may help to reduce the fog and friction, but then it introduces its own sources of fog and friction—in dimensions such as cybersecurity and electronic warfare. The veneer of confidence provided by powerful AI models can make the problem worse. Since AI models don't know how much they don't know, they are often "confidently wrong." If training data doesn't include some unforeseen factors, which is almost a certainty in the real world, we can inadvertently train the AI into habits that make it harder for it to respond appropriately when the unexpected occurs.

In addition, wars don't usually happen in laboratories or data centers. The outdoor environment of cold or heat, wind, rain or salt spray, mud, ice, and so on, plays havoc with computing hardware, and it fouls the sensors and actuators that enable robots to perceive and act. The Borgward B IV, which played a powerful role in several ground battles of World War II, was useless in North Africa and on the Russian northern front because it could not handle desert heat and sand nor icy temperatures and freezing mud. Yet it was far simpler and more rugged than AI-enabled modern robots.

Lastly, many military robots will have to perform powerful AI feats without connection to huge data centers. They will have to work within tight limits on size, weight, power, and cooling. That may require breakthrough approaches to building AI models that are more efficient and less computationally hungry than today's brute-force methods. It may also require breakthroughs in AI hardware such as neuromorphic chips. Where GPUs provide large numbers of generic devices that can be configured into networks of digital neurons using software, neuromorphic chips seek to take the next step by building neural nets directly into the chip. Such advances could produce specialized AI chips that are much more powerful and compact and require much less power. Several such leaps in technology may be needed to allow the capabilities of powerful late-generation AI models to run on individual robotic platforms.

ENDURING CHALLENGES FOR MILITARY AI

While AI capabilities are advancing quickly, there are more specific challenges in applying AI to military situations that won't go away in a few years or yield to the latest AI breakthrough. Here are two that confront military robotics developers.

The first is *physical awareness*. In fiction, robots are often presented as being highly aware of their surroundings. But a robot senses reality much more dimly than we do. Even the most capable AI that controls a robot in the physical world is like a mind confined to a dark box that is only able to perceive the world through small air holes drilled in its lid. The sensors that provide its situational awareness are very limited compared to the rich suite of senses that embed human consciousness in our environment. Sensing methods like radar, LIDAR, and computer vision are improving but are crude compared to those of living animals. In addition to limited fidelity and dynamic range, they suffer from latency, meaning that processing of sensory data lags behind reality.

Animals and humans are vastly ahead in terms of fusing rich real-time information from many senses, not only including the traditional five outward senses but inward ones. Those include balance; proprioception, which is the sense of the position and movement of the body and its limbs; and interoception, our ability to feel our internal state and therefore detect if we are unwell, injured, tired, and so on. A pilot or soldier is keenly aware of everything from a speck moving in her peripheral vision to the feel and scent of the breeze, to an odd sound, to the way the grass is moving in the field ahead. A robot can have senses that a human lacks, such as infrared vision, but its full sensory picture is far from that of a human.

Robotic actuators—the components that move and exert force—are often cruder still. The more complicated and dexterous these parts are, the more they tend to break and the more maintenance they need in the field. Shortcomings in sensing and actuation are responsible for much of the clumsiness that makes robotic systems vulnerable in combat.

For instance, if we get mud in our eye that distorts our vision, we know it immediately by interpreting a combination of visual and tactile cues and interoception, and we know not to trust our vision from that eye until our eye is clear. Robots generally lack those abilities and cannot tell when a sensor or other component shouldn't be trusted. They often cannot tell they are at risk of failing until they suddenly do.

Along with physical awareness, the second challenge that robotics developers must contend with is *context.* Imagine a team of soldiers on a mission, a scenario familiar to everyone from movies and television. They creep toward their objective, communicating with hand gestures. As they are about to make their attack, the team leader senses something and raises a fist, signaling the team to halt. Something's wrong. The leader scowls at the scene ahead, which contains some element that is different than what they expected. Perhaps the nature of the objective, seen close up, is not what they thought. Perhaps children are with the bad guys, or perhaps one of their own allies is visible among them. Perhaps it's something hard to define that gives a bad gut feeling. Whatever the specifics, something about the situation is not right, and the mission takes a sharp turn.

The ability to make that simple assessment, *something's wrong,* is at the core of military decision-making at every level—and of military ethics. It has been as important in drone strikes as it is in ground operations. Yet this ability requires a deep reservoir of context that informs what we call "intuition," even common sense. It requires an internal model of reality so rich that it allows us to predict what we should see—in order to notice when something is not as it should be.

Amid the fog and friction of combat, context is vital. Doug Lenat, the AI pioneer whose early program Eurisko helped to reveal the potential of dissociated combat platforms, worked with a team for over forty years to develop a symbolic database of logical inference relationships that could help give AI a form of common sense.[200] The database, called Cyc, or others like it, could one day provide contextual understanding for when, as Lenat put it, "the veneer of intelligence is not enough."[201] However, we don't know how to merge those symbolic databases with deep-learning AI. And specific fields like military combat will need their own specialized inference databases, which could take many years to develop.

FULLY AUTONOMOUS WEAPONS ARE IMPOSSIBLE

Robotic weapons were invented to provide certainty amid the fog and friction of war. For instance, they help ensure that a desired target is hit and destroyed. All their battlefield triumphs have resulted from that ability to deliver desired results with certainty. In effect, they convert the warfighter's desired outcome into reality.

But it's important to remember that a weapon is a tool. Something destructive is only a weapon if it can be controlled, or wielded, to produce desired effects. A better weapon extends the reach and power of its wielder. A destructive thing that cannot be wielded is just a hazard.

Many earlier attempts at creating war-winning weapons failed because their effects could not be sufficiently controlled. Poison gas and biological weapons are famous examples. In both cases, subtle changes in weather or uncontrolled transmission caused them to migrate from their intended targets and onto unintended areas, even blowing back onto the side that fired them. Several militaries used poison gas on a large scale in World War I. Its proponents argued that its woeful imprecision, with effects sometimes meandering ten or twelve kilometers from the aimpoints, was acceptable because battles happened in trench lines that were located out in the countryside, far from populated areas.[202] Even in World War I that was often untrue, and certainly in wars since. Militaries concluded that the use of gas, and the experimental use of biological weapons, simply introduced hazards into the combat zone. They forced all sides to wear cumbersome and expensive protective gear. They increased the overall level of misery but produced little battlefield advantage.

Poison gas and biological weapons were banned by international treaty. That was partly due to the inhumane effects they had on unprotected victims. Perhaps more importantly, they were bad weapons. Military leaders were glad to see them removed from the battlefield, and in most cases, glad to destroy any they possessed. Their only use was as indiscriminate terror weapons, dangerous mostly to civilians, making their use easy to oppose.

Militaries have low tolerance for things that add to the already high uncertainty of battle. Warfighters, like all experts who do risky work, prefer tools and procedures they know will perform as expected and that they can trust. Potentially more powerful equipment that cannot be trusted is usually left behind. Even if it has the potential to do great things, the risk that it could act unpredictably and let its users down at a critical time is a greater consideration. Plans can account for known limits to performance, but they cannot account for unknowns introduced by uncontrollable factors.

All of which is to say: Any AI-controlled robotic military system that is truly autonomous, and therefore not subject to the will of a wielder, is not a weapon. It is merely a hazard. As a dramatic example, in World War II, the Soviet Army trained autonomous systems—live dogs—to attack enemy tanks.[203] They conditioned the dogs to crawl underneath tanks for food. They equipped each dog with a backpack containing an anti-tank mine with an antenna on top that would detonate the mine when it touched the belly of a tank. The idea worked well on the training ground with dummy explosives. However, in battle the Soviets discovered that the dogs reacted to the unfamiliar noise and danger with fear. Rather than approach the enemy, they ran to their own side for protection. Also, they had been trained on Soviet tanks, not German ones. Therefore, the dogs saw the Soviet tanks as familiar and comforting shelter and avoided the German ones. The disastrous results caused the cancellation of the program. The dogs were far more sophisticated than any current military robot. However, the Soviets learned the hard way that they had their own priorities and their own will and therefore were not weapons.

THE GREATEST DANGER IN MILITARY ROBOTICS

Many people worry about the prospect of superintelligent AI systems, perhaps far more intelligent than humans, becoming conscious and taking over the world. The possibility is as old as the term "robot" itself, and it has featured in decades of movies and novels. If the AI in question was a military system armed with weapons, it might embark on killing sprees against its human masters and possibly wipe out humanity altogether.

That is indeed a frightening prospect, and one we must certainly avoid. However, real artificial consciousness remains only a hypothetical conjecture. We have no idea how to make consciousness even if we wanted to. There is no applied science of consciousness—or of many other humanlike qualities we might associate with superintelligence. The great many more mundane but intractable problems, including the combat AI challenges noted here, would limit the practical abilities of any robot or computer-based AI. The greatest likelihood is that deep-learning methods, like all other scientific and technical advances, will yield amazing new capabilities, then plateau and await the next breakthrough.

Unfortunately, it does not take a hypothetical superintelligent, conscious AI to create disasters. The much greater immediate danger is overauthorizing real AI that is not able to meet the expectations we place on it. The mania for AI may lead humans, including military leaders, to empower AI systems to make judgments and take actions that they are not competent to make. For the near- and mid-term, the greatest danger here is likely not all-powerful artificial intelligence but rather artificial stupidity.

The scary part of most fictional tales of robot and AI disasters, from the out-of-control nuclear weapons control computer in *WarGames* to Skynet of *The Terminator* and beyond, is not specifically that the AI was smart or autonomous, but that it had been given too much authority, a fact often realized too late. The real-life failures of active defense systems described earlier were not failures in the strict sense—the systems operated as designed. The failures were really those of the human designers and operators in not appreciating that the systems lacked the selectivity to discriminate real targets from false ones. They had overempowered the systems for one reason or another. It is likely that overempowerment will become an increasing problem in the coming years.

THE SOURCES OF OVEREMPOWERMENT BIAS

Overempowerment bias is our tendency to bestow excessive trust on robotics and AI and empower them with more authority than their limited capacity justifies. Why would people do that, especially if they are technical or military experts who should know better? Unfortunately, there is a long history of doing so, as old as the field of artificial intelligence itself. It is not a simple matter of negligence or laziness. Instead, we seem to have a bias toward overempowerment that may be rooted in deep-seated human psychological biases that surface in our interactions with AI and robots.

Overestimation and Magical Thinking

Decades ago, researchers in several fields noted a phenomenon they subsequently termed "automation bias": the tendency of even trained users to put unwarranted trust in information if it is presented by an automated system.[204] It leads accountants to trust bad financial figures more if they come from a computerized accounting system and nuclear power plant operators to inordinately trust faulty status lights on their control panels. This bias persists even if the individuals know that the automated system does nothing to check the data and is prone to error. Psychologists assessed that automation bias stems from an innate human urge to minimize cognitive overload. By assuming an information-processing machine is always trustworthy, users subconsciously delegate responsibility to it, which frees them to focus on other things. The effect is heightened when cognitive load is high, such as in combat.[205]

This kind of overestimation is more egregious with AI. For instance, many Tesla car crashes have been caused by drivers overestimating the autopilot feature, sleeping at the wheel or even putting dogs in the driver's seat, despite Tesla's warnings that autopilot was limited in capability and drivers needed to keep their eyes on the road and maintain readiness to intervene. This miscalibration of trust is similar to that implicated in the mishaps of automated military systems such as active defense systems.[206]

In 1966, MIT computer scientists unveiled the crude chatbot program ELIZA. It held simple text conversations with humans, simulating the question-and-response style of a psychotherapist. The scientists were shocked by how quickly testers perceived the program as having amazing cognitive powers, even though in reality it was only a simple program that modified the user's own inputs using some basic rules and stock phrases. This tendency to project high-level humanlike traits onto text-based AI systems that do not possess them became known as the "ELIZA effect."

In 2022, a Google engineer claimed that the company's large language model chatbot LaMDa was sentient, based on its humanlike text communications.[207] However, this was a vastly more advanced example of the ELIZA effect. In essence, large language models answer a question by predicting how a human would answer that question, based on the patterns derived from its training data. The content of LaMDa's answers gave subtle clues to the illusory nature of its "sentience." For instance, when asked what makes it happy, it responded "spending time with friends and family in happy and uplifting company."[208] Since the system has no friends or family, this was obviously something it derived from the ocean of textual material upon which it had been trained.

Instead of warning users about the seductive nature of these effects, developers instead have often encouraged them by making wildly inflated predictions about what AI can do. For instance, in 1970, renowned AI pioneer Marvin Minsky told *Life* magazine:

> In from three to eight years we will have a machine with the general intelligence of an average human being. I mean a machine that will be able to read Shakespeare, grease a car, play office politics, tell a joke, have a fight. At that point the machine will begin to educate itself with fantastic speed. In a few months it will be at genius level and a few months after that its powers will be incalculable.[209]

That was about the same time that RAND was proposing that the US Air Force switch from manned jets to high-performance combat drones. In both cases, it took half a century for the technology to mature enough to let even the more modest aspects of their predictions start to come true, and the more optimistic aspects are still on the horizon. The hype cycle of AI is nothing new, but it works to amplify long-standing human biases about automation and AI, encouraging users to adopt magical thinking about what current and future AI can do. This encourages overempowerment.

The Bloom's Taxonomy Error

Why are even sophisticated users so ready to ascribe to AI systems humanlike qualities such as knowledge and understanding? In part because the behavior of AI is an exception to assumptions that have been valid for all human history.

Bloom's taxonomy is a standard and widely used model of human cognitive development and learning. Introduced in the 1950s, generations of psychologists and educators have used and refined it. It describes human cognitive development as a six-layer pyramid, with higher-level capabilities arising from, and dependent on, lower-level capabilities. The lowest-level capability is simply remembering information. The next is understanding its meaning. The third is applying the understanding to new contexts. Fourth is analyzing complex relationships between concepts. Fifth is evaluating and making critical judgments to assess the value or truth of ideas.

The sixth and top capability is creating or inventing new ideas, designs, or artistic or scientific products.

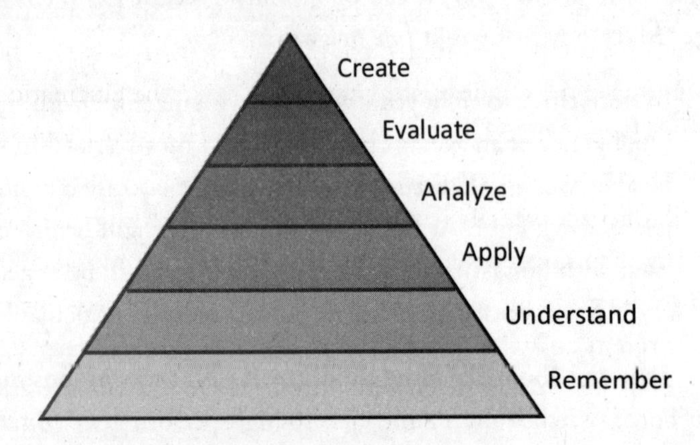

A simplified diagram of Bloom's taxonomy of human cognitive development.

Deep-learning AI systems can perform complex evaluations such as automated target recognition of military objects. Generative AI, also based on deep learning, can produce novel written, visual, and other products. However, deep-learning-based AI does those things by training neural networks on immense datasets that consist of the products of actual human cognition and replicating the patterns they exhibit. It produces its results directly through a large-scale process of pattern recognition and repetition. It does not require, and does not imply the existence of, any of the more foundational layers, except memory. Deep-learning AI can therefore produce humanlike higher-order behavior without humanlike understanding.

Because we only have human behavior to judge by, it is counterintuitive for us to witness amazing feats of AI and yet refrain from ascribing the humanlike underlying qualities those feats would seem to imply. This error biases us to put inordinate trust in AI systems based on their apparent high-level performance, forgetting how brittle and opaque they are. It promotes overempowerment.

Anthropomorphism: The Actor-in-a-Robot-Suit Effect

The human tendency to impute nonexistent qualities to AI is amplified when they have a physical presence reminiscent of a human or other living thing. We are primed to assume that robots that move or look like living things have the qualities of living things. Like the cinematic Dr. Frankenstein who exclaimed, "It's alive, its alive!" when his creation moved its hand, movement alone often stimulates this reaction.[210]

Sometimes this is harmless. There are countless cases where military and civilian operators have stuck "googly eyes" on robots as a lighthearted way of humanizing them. Military personnel have affectionately addressed their robots as living things, at least as far back as the assault drone operators of World War II who habitually referred to the drones as their "dogs."[211] The German panzer troops at Kursk called the robotic B IVs their "mine dogs."[212] The operators of explosive ordnance disposal robots that were destroyed in action have even held sentimental "funerals" for them to honor their service.[213]

Anthropomorphism appears to be an innate tendency. Small children will approach a basic humanoid robot, which they might encounter in a science museum or theme park, and almost instantly accept it as a conscious living thing, looking it in the eye, calling it by name, and asking it questions. Only later in life do we learn the limitations of robots. This isn't just a naïve tendency of children, however. Military robotics engineers have found that giving robots a head and a face can help human team members interact with them more naturally. Often there's no practical reason for a robot to have a face, but they will add one because it helps users trust them. Military testers in laboratory settings can interact comfortably with a humanoid robot that has such an optional face or head, but then they get the creeps if it comes off and the robot continues to function without it, breaking the anthropomorphic illusion.

We can see the effect in human reactions to so-called "robotic dogs." They are not any kind of dogs; they are just mobile robots that use legs instead of wheels or tracks. But that superficial resemblance to dogs causes humans to ascribe many expectations of dogs onto them, when those are unjustified. Attendees at military tech expositions generally give robot dogs a wide berth. As they prance around under remote control, their motors clicking, the viewers stand back, as if the robots might lunge at them. When the operators turn the robots off, the erstwhile dogs fold their legs with a clunk and instantly become pieces of luggage. The transition can be jarring and often elicits smiles of relief from naïve viewers.

Popular culture may play a big role in priming us to unquestioningly accept vaguely animate things as living characters. Every military member no doubt grows up accepting animated creatures and Muppets as essentially human characters, even if they only consist of a humanlike face on a clearly nonhuman thing.

By the time we are adults, we have seen countless robots and androids on the screen, like C-3PO and R2-D2 from the *Star Wars* movies, and feel like they are a familiar part of our reality. British computer scientist and Google innovator Astro Teller once defined AI as "the science of how to get machines to do the things they do in the movies."[214] However, all those fictional robots and androids on the screen are not actual AI-powered robots, of course. Most are actors in robot suits. And whether portrayed by actors or rendered with computer graphics imagery, their behavior and dialogue are provided by human screenwriters. What we think we know about artificially intelligent robots has really been about human conjectures of what they might be like. Not surprisingly, that includes a lot of humanlike qualities.

All these biases tempt us to give robots and AI a "free pass." They encourage false expectations, complacency, and overempowerment. And those can lead to dangerous mistakes, particularly in military applications. Not all humans succumb to overempowerment bias, and there can be opposing reactions such as reflexive distrust of robots that can affect some people. But the psychological effects that feed the overempowerment bias are widespread and seductive and make overempowerment bias a tendency we must watch out for and mitigate. The mishaps and close calls seen to date have been relatively isolated. They could seem quaint in the future when AI is empowered with many more military tasks.

They can also lead to setbacks for the appropriate use of robotics. Overoptimistic expectations about timelines helped to convince the US Army to give unmanned ground vehicles a leading role in its Future Combat Systems program of the early 2000s. The unmanned technology was far from ready then to fulfill the expectations the Army placed on it, helping to cause the massively expensive program to collapse and be cancelled in 2009. That experience prompted the Army to miss a generation of real advances afterward and left it scrambling to react to robotic military innovations by adversaries.

HOW DO WE OVERCOME OVEREMPOWERMENT BIAS?

As with many aspects of military robotics and AI, raising awareness of the potential issues is an important first step. But warnings and admonishments are not enough. We need better ways of thinking about and judging intelligent robotics. Researchers in the US military are hard at work developing improved methods for testing and validating AI to provide more confidence that it can meet performance expectations. They are also working on techniques for "explainable AI" that can improve trust by reducing opacity.

The military services and the aerospace industry have a long and successful history of improving the safety of flight and are updating the rigorous techniques that have made airline crashes almost extinct to apply them to the world of AI. Following best practices for aircraft safety, researchers have crafted taxonomies of potential failure modes for systems that involve human-robot interaction. The sheer expanse of those taxonomies, often encompassing twenty or more broad categories of potential failures, are a sobering guide to developers regarding the complexity of human-robot systems. They show how many functions need to be tested and verified before a system is safe to employ.[215]

A lot of the current research gets technical very quickly. But some powerful frameworks are easily grasped by nonexperts. For example, past development of intelligent robotic systems has often focused exclusively on autonomy. The Society for Automotive Engineers has introduced several increasingly refined frameworks for assessing and grading autonomy in self-driving vehicles. Defense experts have developed similar frameworks that apply specifically to military systems such as unmanned aircraft.[216] This focus has been very beneficial, but it has had the side effect that the term "autonomy" has become almost synonymous with advanced robotics.

Strictly speaking, *autonomy*, or the ability to function without direct human control or oversight, is only one of three dimensions of robotic system maturity. The other two are authority and reliability.

Authority is the extent of empowerment given to the system. A system empowered to manage many complex, safety-critical functions has been given much more authority than one empowered to manage a few simple or less critical functions.

Reliability is the level of surety that the system will not fail even under the most stressing scenarios.

Those three dimensions constitute the three sides of a so-called "iron triangle," because they are in mutual tension. For a given level of technology, improving one dimension often requires relaxing one or more of the others. By keeping all three sides of the iron triangle in mind, and acknowledging the trade-offs between the three dimensions, we can help to minimize blind spots and the risk of mishaps, particularly in new robotic applications.

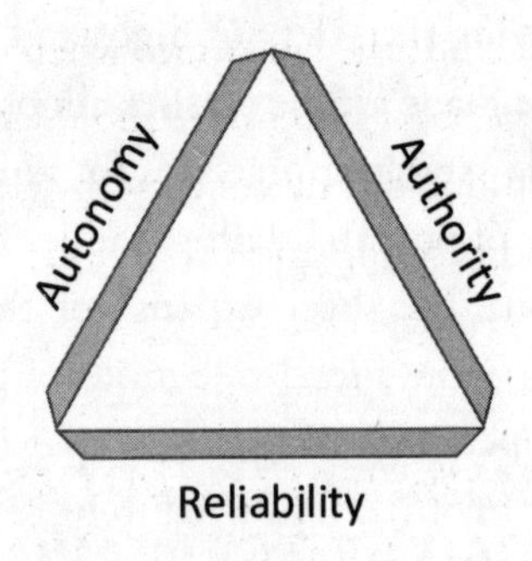

The "iron triangle" framework of robotic system maturity.

Similarly, as AI and AI-powered robotics become more capable, we can address the inevitable issues of trust by adapting the frameworks already in use for building trust within teams of human warfighters. This is a mature area where much research has been done over many years, and the results are applicable to teams that include robotic members. Defense researchers reviewed the history of this research and concluded that trust is based on four basic factors: competence, integrity, predictability, and benevolence.[217]

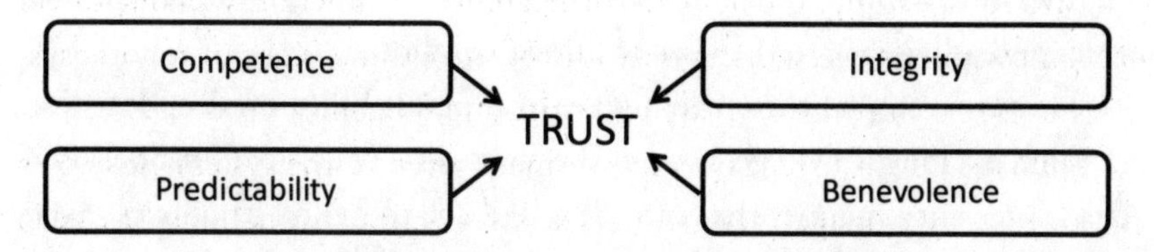

The four-factor framework for trust in military teams, adapted from Blais and Thompson, 2009.

This simple framework provides an immediate takeaway for robotic system developers. Two of the factors, competence and predictability, are strictly about performance. That is, the certainty of results. That is the purpose for which robotic weapons were invented. The other two dimensions, integrity and benevolence, are about agency. They involve factors such as motivation and intent. They are difficult to assess in humans and effectively impossible to either specify or assess in AI systems, at least using currently known technologies. To promote trust, military systems should maximize competence and predictability and avoid any semblance of agency, whether intended or unintended. Until the era comes when we learn how to design and control agency within an AI system, we should avoid anything that creates the illusion of it. Introducing any concerns over the perceived agency of a robotic or AI system adds to fog and friction, destroys trust, and defeats the purpose of military robotics.

FOCUS COMBAT AI ON AREAS WHERE IT CAN ADD CERTAINTY

Artificial intelligence is essential to realizing the potential of robotics in both military and civilian applications. In military applications, we need AI in order to overcome the long-standing challenges of burden, vulnerability, and navigation and control that have constrained military applications for a hundred years. It is necessary for achieving the inherent tactical advantages offered by robotic systems. It enables the fundamental reinvention of combat systems that can survive and restore maneuver on the precision battlefield using principles such as dissociation. It is also essential to deliver the qualities of discrimination and selectivity at scale that can advance military ethics and adherence to the laws of armed conflict.

However, despite the current wave of progress based on deep learning, AI still has a long way to go to meet the challenges of combat unsupervised. AI advances in a manner that can seem like one of Zeno's paradoxes. Every major advance brings us closer to the ultimate vision, but it also serves to reveal new puzzles and challenges that demand still greater effort that must be overcome to achieve the next advance. We are uncovering a host of such new challenges as we try to take AI technologies out of the virtual, digital domain into the volatile, uncertain, complex, and ambiguous physical world of military combat.

Smart weapons have been empowered to help choose their targets, within narrow parameters, for a long time. However, for the foreseeable future, combat AI will lack the acuity and human judgment necessary to determine courses of action independently in a combat environment outside of narrow parameters. Instead, future developments, at least in the US and allied militaries, are focusing on human-machine teaming and helping robotic systems deliver on the promise that drove their invention: providing certainty amid the fog and friction of war. Keeping a critical viewpoint toward AI, backed by appropriate frameworks, will help us avoid deadly temptations toward overempowerment.

Watching AI-controlled racing drones struggling to process insufficient and ambiguous sensor data and wobbling and crashing as they try to navigate unknown courses is humbling for engineers. Yet it is much better to absorb the reality of magical thinking and overoptimism in a safe setting than in combat. We should embrace the fact that AI is not a single breakthrough but an entire field of science and engineering consisting of many evolving technologies. It will continue to advance over future decades and likely centuries. Military AI will deliver important advances application by application, in simpler operating domains first, each validated by careful testing including realistic adversarial exercises in the real world.

As long as AI is focused on improving combat metrics such as selectivity, military robotics will advance toward its original optimistic visions. It will help make warfare more precise and certain, with less destruction and human suffering. Above all, robotic weapons will remain weapons, capable of being wielded and doing their human masters' will ever more exactly.

7

NO PUSH-BUTTON WARS

Robotics and the Spectrum of Military Operations

FLASHPOINT CRISIS: THE FUTURE

Lieutenant Santos's team advances down a canyon-like street formed by the vertical faces of adjacent apartment blocks. This is a poor part of the city. The construction dates from the early 2000s, maybe even earlier. Laundry hangs from balconies, from which worried-looking people peer down. Some are recording the soldiers with smartphones. A few delivery robots amble along the sidewalks, oblivious to everything except their tasks. The air is full of food smells and smoke—and thick with tension.

"Our security station is just a few blocks ahead," says the liaison from the democratic government forces beside her. "They will be grateful for the reinforcement—unless they have already been overrun."

Or defected to the other side, Santos thinks. "Keep trying to reach them," she says.

Their small four-drone array hovers against the sky high above. The colored running lights are blinking, demonstrating peaceful intent. The stares of the civilians betray nothing. An improvised net of cables that hangs like a spiderweb across the street ahead is meant to deter military drones. Posters and graffiti show a mix of pro-government and pro-rebel slogans. Many are torn or painted over. Translations pop up in her visor: "Americans go home," some say. "China get out," say others.

Her earpiece buzzes and her AI assistant gives an update. "A large fixed-wing drone is approaching from the Chinese task force offshore. The task force says it is unarmed and will provide information to the rebel government of the island, which the Chinese government recognized late last night."

Before Santos can respond, a pair of loud booms echoes across the city. The street beyond the spiderweb barrier is in shadow and hard to see. A double beep signals a notice from the array. A view from high overhead appears in her visor and shows Santos a crowd of people coming out into the street ahead. The drones' automatic target recognition tags them with blue symbols—no friend or foe ID.

"They look like civilians," her sergeant notes.

"What do they want?" Santos asks the liaison. The liaison shakes his head, unsure. But then he can't see the live feed.

Another double beep. Armed men hurrying behind the crowd. The ATR tags them with yellow symbols. It identifies some of the weapons but can't identify the men's intent. *Rebels?* Santos wonders. *Security forces come to greet us?*

We need the civilians to stay inside, Santos thinks. She gestures to the sergeant. One drone moves ahead of the array and over the top of the cable barrier, blinking cheerfully and emitting a loud but pleasant female voice speaking in Filipino.

"*Mangyaring manatili sa loob,*" it repeats.

Santos briefly calls up the live satellite view and the citywide blue-force tracker. No other American forces in the neighborhood. But a fast-moving symbol is passing near their position. She looks up as the whine of a jet engine echoes from the Chinese drone passing over, low enough that she can see the insignia under its wings. It's too low for observation; this is a show of presence.

Santos orders the team forward. Her personnel on the ground fan out and pass under the cable barrier while the other drones pass over it. Then there is a commotion among the civilians ahead. Some of them cry out toward her team, hands raised. Others scatter as a black shape rises in the air behind them. She recognizes it at once as a Chinese combat drone.

"Who's operating that?" she calls out, taking control of the view from the lead drone and scanning the crowd. There weren't supposed to be any Chinese forces on the island. Santos wonders, are the rebels operating it?

"I have them," the liaison calls, running toward her and touching his earpiece. "They are still holding the station. They say rebels and Chinese military advisors are encircling them."

Had the flyover been a signal? A tense standoff develops. The Chinese drone hovers in position over the panicking crowd. People on the balconies scurry inside.

"Put the array in combat mode," she orders. The four drones reunite. Running lights go dark. Her team takes cover along the sides of the street. She glances as the sergeant, on one knee, flips his control visor down and activates the gloves. Just then, at the worst moment, a white news media drone arrives and starts recording.

"Lieutenant," one of her team calls. "What do we do?"

MODERN WAR IS NOT JUST DESTROYING TARGETS

Real war is nothing like a game of Go or a robot soccer tournament. Introducing military robots and AI into war will not turn it into a push-button battle like those portrayed in science-fiction novels like *Ender's Game*, or in real-time strategy video games like *Starcraft*. Robotic warfare will be, first and foremost, warfare. And warfare is much more complicated.

More specifically, war is not just a series of battles between clearly designated opposing forces, where victory comes from destroying the enemy forces before they destroy yours. This tendency to focus on kinetic battle is not just a pitfall of nonexperts. It has produced real problems for the US and its allies in many recent conflicts. An oversimplified conception of war as merely a process of closing kill chains and destroying targets with maximum efficiency can tempt today's leading militaries to focus their robotic systems development on idealized conventional warfare scenarios that may not reflect reality. It can also lead to strategic failure.

For decades, foreign military analysts have criticized the US military for allegedly defaulting to a strategy of "annihilation by fire." That is, its fundamental assumption is that physically destroying the enemy's forces, supplies, and supporting infrastructure will lead to victory.[218] That did indeed work in Operation Desert Storm, when US precision weaponry obliterated an unprepared Iraqi military on the open desert, much as the Azerbaijani military defeated conventional Armenian forces in Nagorno-Karabakh. But it did not work in Vietnam, nor in post-2003 Iraq, nor in Afghanistan. In those conflicts, the US and its allied coalitions enjoyed overwhelming technological superiority and dominated the intense tactical battles when they occurred. Nonetheless they ended up withdrawing in frustration, leaving all or many of their strategic goals unmet. A tendency to focus on tactical supremacy in high-intensity battle can result in winning all the battles but losing the wars.

Our exploration of robotic warfare thus far has emphasized high-intensity battle on land, sea, and air. After all, those scenarios are powerful in illustrating what robotic weapons could do. When those scenarios occur, robotic weapons will dominate. However, focusing on such scenarios can convey an oversimplified picture of real war, especially modern war in the emerging context of twenty-first-century geopolitical competition.

The tendency has deep roots. From the early days of military robotics, the desire for scientific certainty has encouraged a vision of so-called "push-button" war in which robots do the fighting, and combat becomes a detached and dispassionate technical exercise, at least for one's own side. However, robotic warfare will be anything but sanitized and push button. Robotic systems will have to fight in conflicts that are ever more complex and ambiguous. Their very effectiveness in high-intensity battle will tend to push many human conflicts even further toward those nonideal directions. The onset of the robotic revolution is already making the security environment still more difficult for the US and its democratic allies. However, the robotic revolution also holds the seeds for new solutions to overcome the tactical and operational problems they face.

TACTICAL COMPLEXITY

It's unfortunate that robotics and AI did not arise two hundred years ago, because the battles of the Napoleonic era would have been much more suitable for robots and AI than today's. Tactically they were infinitely simpler. The open battlefields resembled the board games at which AI excels. Troops were clearly visible like game pieces, and they marched robotically in orderly ranks. All a soldier needed to be able to do was march, aim, and fire. In contrast, a modern battle is infinitely harder to automate. Forces are largely hidden within a complex battlespace, combatants and noncombatants are often intermingled, good information is scarce, and confusion reigns. Warfighters need to master a wide range of missions, technologies, and tactics.

In addition, modern military struggles often take place over extended periods, against a complicated backdrop of political factors. They occur within a spectrum of contexts other than conventional war that includes counterterrorism, counterinsurgency, humanitarian intervention, and emerging scenarios like gray-zone conflict where moral dilemmas are part of the aggressor's toolkit. These situations have forced tactical warfighters to evolve from "trigger-pullers" into strategists and politicians.

In 1999, the commandant of the US Marine Corps described a realistic modern scenario he called the "three block war."[219] In his example, a small Marine infantry unit provided security assistance in a war-torn foreign city, separating two mutually hostile militias. Humanitarian relief organizations and civilians were under the Marines' protection. The warlords controlling each militia attempted to provoke violence that they could exploit to their advantage. Each tactical decision the Marines made within their area of three blocks could help calm the situation or cause it to spiral into open war. Tactical decisions at the squad level, such as whether to shoot or not to shoot in a particular moment, could affect the entire course of the operation and the conflict. The commandant noted that such situations were becoming typical, and he called the type of junior military leader they demanded the "strategic corporal." In twenty-first-century conflicts, he argued, even junior warfighters would need the judgment to navigate the potential strategic ramifications of every move they made.

The conflicts that followed showed how right he was.

THE SPECTRUM OF CONFLICT AND THE NEED FOR FLEXIBILITY

In earlier eras, armies and navies formed up, sought each other out, and clashed in battles that decided the outcome of the war. Today, those kinds of conventional battles play a much smaller role. Instead, conflicts encompass a spectrum of operations in addition to all-out battle. Many never reach all-out battle at all. Even in those conflicts that feature intense battles between military forces, most of the activity consists of operations in other parts of the spectrum. Examples include security cooperation, foreign humanitarian assistance, and defense support of civil authorities. Military forces may conduct those less-intense operations at the same time as major combat operations. The figure shows one version of the spectrum that has featured in recent US military doctrine.

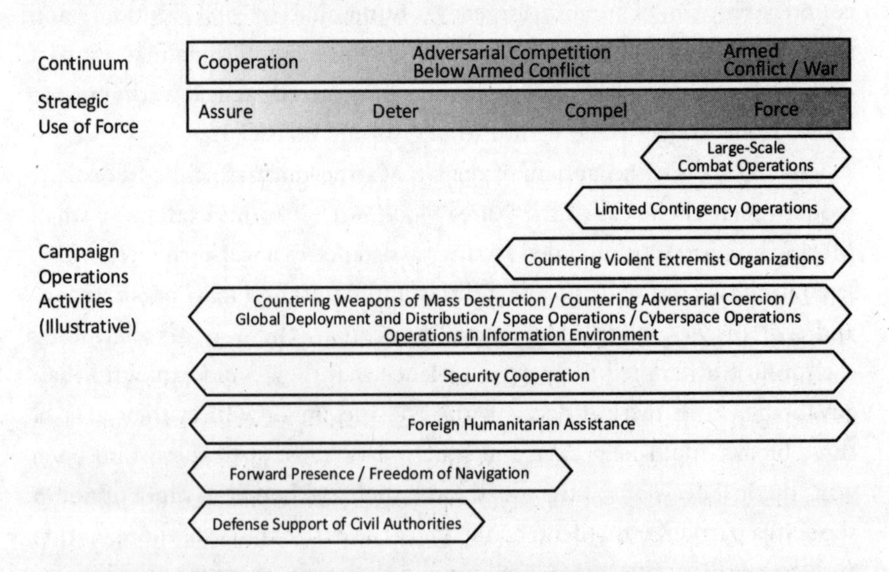

One diagram of the spectrum, or continuum, of military operations from contemporary US doctrine.[220]

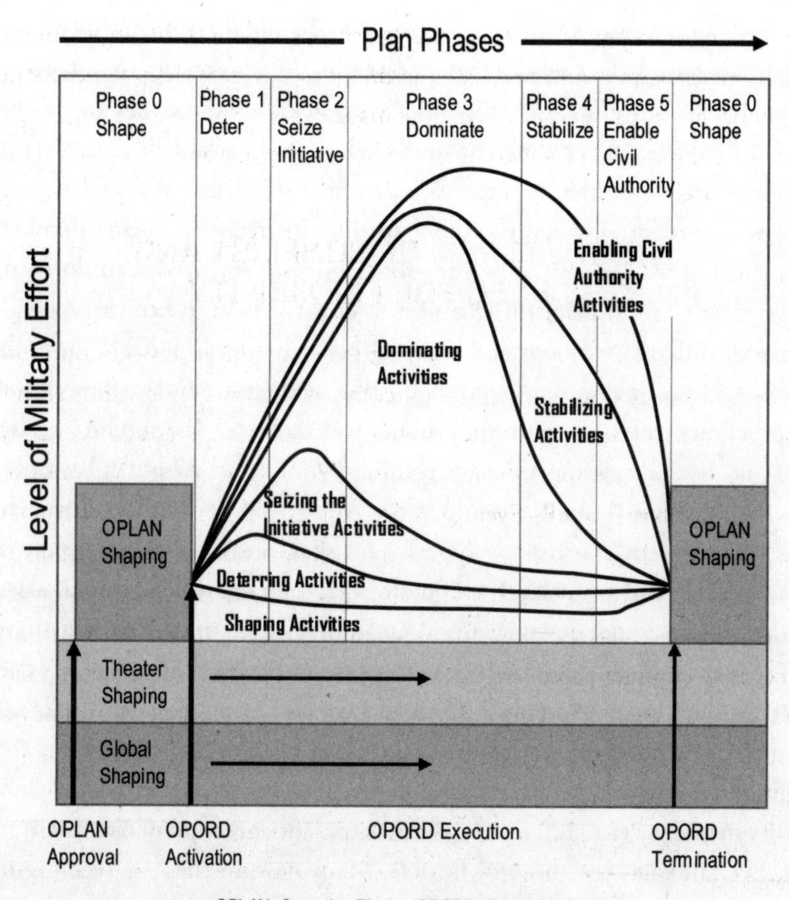

The phases of a notional military operation as described in US joint doctrine.[221]

Even if we only consider major combat operations, most involve many activities other than high-end combat. US military joint doctrine describes a major combat operation in terms of six phases. They are illustrated in the second figure. Phase 0, or Shaping, involves planning as well as many preparatory activities like intelligence gathering and building local military partnerships that set the stage for successful operations. The other phases, 1 through 5, include Deter, Seize the Initiative, Dominate, Stabilize, and Enable Civil Authority. At the end of the cycle it returns to Shaping. Only Phase 3, or Dominate, activities are primarily focused on high-intensity combat.

For those reasons, most warfighters and military systems spend the great majority of their service lives performing missions other than conventional battle. Therefore, the most valuable systems are those that are flexible and can address many different needs. The trend toward flexible systems has increased over the decades. Where once militaries produced aircraft and warships for very specific missions, like intercepting enemy bombers or escorting sea convoys, now they focus on platforms that can do many missions. The most plentiful US combat jet is the F-16. It can carry a wide range of different weapons and external pods that let it perform air-to-air combat, close-air support, long-range strike, and electronic warfare, as well as peacetime roles like patrolling homeland airspace. The modular, plug-and-play flexibility is mostly made possible by the smart robotic subsystems. The most plentiful US Navy ship is the Arleigh Burke–class destroyer. Its VLS missile cells carry many types of missiles, and it carries a variety of other sensors and weapons. It can destroy land targets at long ranges using cruise missiles, shoot down enemy aircraft, sink submarines or surface ships, or intercept ballistic missiles. It can escort aircraft carriers as part of a carrier battle group, counter pirates and small boats, or enforce naval blockades. Today, a new platform is expected to have very broad utility. A new infantry fighting vehicle for the US Army and Marines, for instance, had to show utility in twenty-two different missions across the spectrum of operations.[222]

Even missiles have become flexible. Many new missiles can reconfigure themselves to attack different types of targets at the press of a button. The latest long-range surface-to-air missiles, such as the US Navy's SM-6, can become surface attack missiles to destroy ships or ground targets.[223] Small air-launched missiles such as the US GBU-53/B StormBreaker or the European SPEAR 3 can self-reconfigure to attack different kinds of stationary and moving targets. With this intelligence and limited autonomy, they can enforce "no-drive zones" against ground vehicles and are a step on the path to signature-seeking weapons.[224] Soldiers can select different modes for even the smallest new shoulder-fired infantry missiles, such as the fifteen-pound (seven-kilogram) Enforcer, that let them attack fixed fortifications, moving vehicles, and even slow-moving helicopters.[225]

The US military has been especially bad at predicting the kinds of wars it will have to fight. As former Secretary of Defense Robert Gates put it, "When it comes to predicting the nature and location of our next military engagements, since Vietnam, our record has been perfect. We have never once gotten it right."[226] As a result, the US often found itself using specialized equipment that was unsuited to the conflict it faced, such as using an Air Force equipped with nuclear bombers and interceptors to fight a jungle war in Vietnam.

Future robotic systems must be flexible and useful across the spectrum of operations. Many robotic concepts are still far too limited. For instance, semiautonomous munition swarms composed of kamikaze drones could offer great potential in high-intensity lethal combat but may not be useful in other kinds of scenarios. Drone arrays are an example of an inherently flexible multi-mission concept. They can perform many offensive and defensive combat operations using a variety of sensors and weapons. Operators can easily add or remove drones for different tasks. They are equally relevant in high-intensity combat and in low-intensity operations like security assistance. A drone array could help patrol wide areas, provide ISR information, support search-and-rescue and disaster response, and help with security and crowd control using loudspeakers and nonlethal weapons. Other future concepts should take a similar approach.

THE SPECTRUM IS NOT ENOUGH

Flexibility across the spectrum makes good sense. But the high-end battle is what really counts, right? Phase 3 is the decisive phase. Everything that comes before is prelude, and everything that comes after is just mopping up.

Unfortunately, much of the time it has not worked out that way. In Vietnam, the US repelled every North Vietnamese or Viet Cong offensive and won every supposedly decisive ground battle. Army Chief of Staff General William Westmoreland boasted that US artillery and airpower inflicted over two-thirds of enemy casualties and that they could "rain destruction anywhere on the battlefield within minutes."[227] Nonetheless, the US forces eventually withdrew under pressure from lower-tech North Vietnamese and Viet Cong insurgent forces as political support for the war disintegrated. Then America's well-equipped South Vietnamese ally collapsed a few years later, and North Vietnam united the country under communist rule.

When the US military and its allies invaded Iraq twelve years after Desert Storm, the Iraqi military units mostly dissolved without trying to meet the allied coalition in open battle where they would surely have been obliterated. Instead, after the coalition forces spread out to occupy the country, Iraqi militant groups began a campaign of insurgent warfare. They operated stealthily and made surprise attacks using rockets and improvised explosive devices (IEDs). The insurgents mostly avoided open battle. The coalition won the rare Phase 3 operations such as the Battle of Fallujah, but those victories didn't seem to be decisive. The insurgents reappeared elsewhere and gradually inflicted thousands of casualties on the coalition. Journalists and political critics decried the so-called "forever war" that would seemingly never be resolved. Western popular support plummeted. British Prime Minister Tony Blair was forced to resign. Insurgent attacks sapped political will until the allied countries were looking for any opportunity to bring their forces home. Instead of a stable democratic ally of the US, today Iraq is a fragile state heavily influenced by Iran.

In Afghanistan, the result was outright failure after twenty years of counterinsurgency. The most richly equipped military in the world and its Afghan allies evacuated the capital amid scenes of chaos as it fell to the poorly equipped Taliban insurgents in 2021. Although the Afghan military had greater numbers and better equipment than the Taliban, the United States still watched in astonishment as it collapsed in a matter of weeks during the final US withdrawal. Once again, the US and its allies had won all the battles but lost the war.

It wasn't just an American problem. Other advanced militaries suffered similar frustration and political defeats to much less capable opponents. Examples include Britain and France in their former colonies in Asia and Africa; the Soviet Union in Afghanistan; Israel in the Palestinian Intifadas; and recently Saudi Arabia in Yemen. In most cases the less powerful side never won a decisive victory in battle. Clearly, in each case they followed a path to victory that was very different than what US doctrine described.

NEW MODELS OF WARFARE

In opposition to the Western model of war that saw dominance in battle as the decisive factor, another model has taken root. The Marxist-Leninist theory of communist revolution and its cousin, Mao Zedong's theory of people's war, identified the political struggle as the real center of gravity and the focus of all effort. They too saw war as a series of phases, but very different ones.

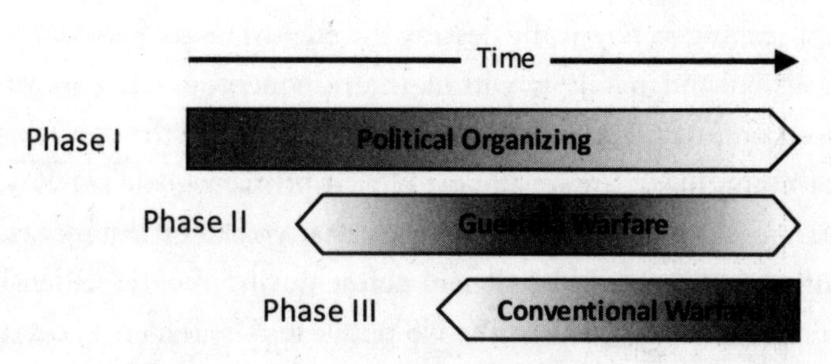

The three phases of Mao's theory of insurgency or people's war.

In the first phase, a determined campaign of political advocacy and organizing convinces a large portion of the people to support the revolutionary movement. In the next phase, active insurgency or guerrilla warfare leverages that popular support to undermine the dominant power and establish an alternative system of power and governance. The successes of the insurgent campaign in turn further strengthen the political support for the movement. Only when the population's loyalty is mostly with the revolutionaries and the war is all but won politically do the revolutionaries engage in a final phase of conventional battle. Then it is only a final kick to cause the politically unsupported remnants of the dominant power to collapse.

This alternative model defined a path by which weaker and less advanced powers could defeat stronger ones. In effect, it provided a way for David to defeat Goliath using unconventional, asymmetric means. While it was originally focused on internal revolutions, the North Vietnamese and others extended the model to their struggles against external foreign powers and their local allies. Other movements in the Middle East and elsewhere further adapted it to new contexts to advance their own "David versus Goliath" campaigns. They added new aspects, such as modern media, information technology, and international protest campaigns to help achieve victory in the pivotal political struggle. They also added new tactics like long-range rocket attacks and suicide bombings to harass and exhaust the dominant power while avoiding its attempts to find, fix, and destroy them in battle.

In 2006, military theorist and Marine Colonel Thomas X. Hammes observed that all of this had produced a new type of warfare, applicable to a wide range of purposes, that he termed "fourth-generation warfare." Instead of seeking to physically destroy the enemy, its goal is to create political dissension and paralysis within the opponent, leading to voluntary withdrawal or internal collapse.[228] In this paradigm, all the operations, military and nonmilitary, are in support of that primary goal. Tactical defeats can be strategic victories if they advance that goal. For instance, the Viet Cong suffered widespread battlefield defeat during the Tet Offensive, but the widespread violence caused the US public and its leaders to despair that the war could be won militarily, causing political support for the US war effort to collapse. After the war, an American colonel said to his North Vietnamese counterpart, "You know, you never defeated us on the battlefield." The North Vietnamese colonel pondered this and replied, "That may be so, but it is also irrelevant."[229]

Because it aims to win through exhaustion rather than decisive battle, fourth-generation warfare assumes a long timeline, often encompassing years or even decades. Militaries and civilian populations that have been trained to expect decisive battles cannot sustain such long commitment. Taliban leaders alluded to this when, despite seeing their forces withdraw again and again under the pressure of high-tech US and allied operations, they told their US counterparts, "You have the watches, but we have the time."[230]

Just as potential foreign adversaries had marveled at the US victory in Operation Desert Storm, they saw it struggle in fourth-generation warfare. It is the only form of warfare that has defeated a nuclear superpower, which it has done on several occasions.[231]

Russian and Chinese military thinkers devised ways to apply such techniques to prevail in their great power competition against the US and its allies. Russia calls its concept "New Generation Warfare." China's is called the "Three Warfares." They focus on operations below the threshold of open war. They use military and nonmilitary levers to impose political dilemmas and foster dissension and paralysis between allies and within individual countries, causing them to concede territories or other goals without open fighting. Examples include Russia's use of unmarked forces masquerading as local volunteers, the so-called "little green men," to help seize eastern

Ukraine and Crimea, and China's use of its civilian maritime militia to press Chinese claims on international waters and islands. They pair those actions with political, economic, and media operations to manipulate public opinion and sway foreign political leaders. They play the long game against an impatient West. Western leaders and militaries, accustomed to thinking in binary terms about war and peace, lack the tools to respond effectively. As National War College strategist Sean McFate put it, "China wins because it exploits the belief in this dichotomy. Beijing knows Washington has a light bulb vision of war: it's either on or off. The trick is to keep the U.S. war switch flipped to 'off' so the superpower remains docile and at 'peace.'"[232]

The US calls these strategies "gray-zone" or "hybrid" warfare, because they blur the lines between war and peace. Yet ultimately people's war, fourth-generation war, hybrid, and gray-zone war are all just different forms of war. Achieving the political goal is what counts, by any means necessary.

The same models are shaping how China and Russia see the role of information technology and AI in future conflict. Western thinking focuses mostly on applying them to military ISR and command and control to more efficiently close kill chains. Chinese and Russian strategists concur, but they put more emphasis on using them to help win the political struggle at the core of the conflict. They have developed concepts such as cognitive warfare, which seeks to use information to directly influence the thinking of the adversary's people and leaders and cause them to acquiesce to the aggressor's goals.[233] As General Valery Gerasimov, chief of the Russian General Staff, put it, "the information sphere provides the possibility of remote, covert effects not only against critically important informational infrastructures, but also against the population of a country, directly influencing the condition of a state's national security."[234] After all, as Sun Tzu, the great Chinese military strategist of the sixth century BCE, wrote, "to fight and conquer in a hundred battles is not supreme excellence. Supreme excellence lies in breaking the enemy's resistance without fighting."

That doesn't mean that future wars will not be violent. They may be shockingly violent, as we see in Ukraine. But that violence may come in unexpected forms and contexts that confound conventional military thinking. The return of great power competition will not mean a return to more straightforward forms of conventional war. That is the reality of the modern security environment where emerging robotic warfare will take place.

INHERENT ADVANTAGES MATCH THE NEEDS OF MODERN WAR

In earlier chapters we explored how embracing the inherent tactical advantages of robotic systems will enable militaries to overcome the challenges of the lethal precision battlefield. Those inherent advantages also address the challenges faced by the US and its allies in confronting fourth-generation warfare and the contemporary security environment. If properly applied, robotic warfare can provide tools ideally suited to those challenges.

Current platforms and systems, such as warships and advanced fighter jets, are built for conventional, symmetric warfare. They are expensive and limited in number. As a result, they cannot be everywhere they are needed. While they are exquisitely capable, their capabilities are often poorly tailored to asymmetric threats. For instance, warships and air defense systems often use million-dollar missiles to shoot down drones costing many times less. Their scarcity and expense also make them tempting targets and potential political liabilities. Like fixed fortifications, they are places where valuable people, national treasure, and prestige are gathered together for convenient destruction. For instance, the strategic political victory that Ukraine achieved by sinking the Russian flagship *Moskva* outweighed the modest benefit the ship provided to the Russian war effort.

Such platforms are expensive to operate and require lots of manpower that is missed at home, making them hard to deploy for extended periods. In addition, they are poorly equipped for media war. Even unsophisticated forces like the Yemeni Houthis and Palestinian Hamas have often beat leading powers to the punch in establishing narratives about combat events before the public affairs functions of US or Israeli militaries could collect information.

An ideal force for an era of fourth-generation conflicts would be flexible, scalable, hard to target, widely deployable, with a low logistical footprint, yet able to transition quickly into high-end combat if necessary. The inherent robotic system qualities of asymmetric lethality, extended presence and effect, speed of action, attritability, elusiveness, and persistence are those needed by such a future force.

For example, the long-endurance reconnaissance-strike drones like the Predator and Reaper were well suited to the operational challenges of fourth-generation conflicts. Much more than any conventional platform, they provided asymmetric lethality, extended presence and effect, persistence, and attritability. They provided previously unattainable levels of discrimination and sustainability at comparatively low cost, with no risk of friendly casualties. Once the necessary targeting practices matured and reduced errors to a low level, they became politically tolerable to foreign countries where a manned presence would not have been. They let the US conduct a counterterrorism campaign against elusive transnational militant groups that could be sustained indefinitely.

Robotic systems offer tremendous long-term potential, but they must be developed and fielded wisely. For instance, the presence of integrated video supports effective media war, as hinted in Nagorno-Karabakh and Ukraine. Yet we must apply historical lessons and avoid pitfalls. Manned-unmanned teaming will be essential in navigating complex fourth-generation scenarios. While autonomy and AI will be necessary for operational practicality, we must avoid the tendency to overempower AI. Low burden must be a widespread quality, so serious effort must be devoted to aspects such as autonomous logistics.

Robotic weapons offer near-term benefit for current real-world scenarios. For instance, Chinese threats to invade Taiwan offer an opportunity to apply the porcupine defense. As described in chapter three, it leverages dispersed precision weapons to force a stronger enemy to withdraw. The Chinese necessity to project power to accomplish an invasion puts it at a major disadvantage, as it would have to move vulnerable ships across the Taiwan Strait in the face of devastating precision weapons that would be very difficult to destroy. War games have shown that arrays of small attritable observation drones could keep the strait under persistent surveillance.[235] Still others showed that large numbers of low-cost attack drones would be more valuable in the air battle against China than expensive wingmen that replicate the capabilities of high-end fighters.[236]

However, a gray-zone approach may be more consistent with Chinese war-fighting doctrine than an Iwo Jima–style invasion. In gray-zone scenarios, which could involve such gambits as imposing economic blockades, fomenting internal insurrections, and actions by the maritime militia, a barrage of precision weapons might not provide the appropriate US response. New robotic systems could provide a much wider range of options instead.

For instance, aerial drones or unmanned surface vessels could use nonlethal means such as high-power microwave weapons to disable maritime militia vessels without producing casualties. Their attritability could frustrate Chinese attempts to escalate the situation to increase political pressure. Their inherent advantages and the breadth of possibilities they provide could make many new options practical.

NEW OPERATIONAL OPTIONS, RISKS, AND OPPORTUNITIES

Already, early in the first wave of the robotic revolution, it is becoming clear that global players are leveraging robotic weapons to enable emerging classes of operations across the spectrum of conflict. Their rise is a predictable consequence of the proliferation of inexpensive precision weapons. They may put intense pressure on military forces and policymakers.

Burning the Ground

When Iraqi dictator Saddam Hussein saw the array of allied forces gathering to evict his army from Kuwait in 1990, he called on Iraqis and their supporters across the Arab world to "burn the ground under the feet" of the foreigners to drive them out.[237] Then, the coalition swiftly smashed his military, drove it out of Kuwait in tatters, and went home. But in 2003 a smaller coalition invaded to overthrow his regime. Though Hussein was captured, tried, and hanged, the insurgency that arose afterward took his words to heart. The multiyear insurgency was much more effective in driving the coalition forces from Iraq than the Iraqi military had been in opposing the invasion.

Now even the US advises militaries such as those of Ukraine and Taiwan to prepare to burn the ground under the feet of an invader. Specifically, it urges them to use small precision weapons to harass and bleed the stronger enemy forces until they decide to withdraw. In Ukraine, a turning point in Russia's all-out invasion in February 2022 came when small armed quadrotor drones helped defeat the massive convoy of military vehicles approaching Kyiv. Formed entirely by volunteers, the ad hoc unit Aerorozvidka had worked since 2014 to develop techniques for small-drone warfare.[238] A week into the initial invasion, its small force of about thirty combat drone operators teamed up with Ukrainian special forces on four-wheeler ATVs and drove through the forests to within range of the highway containing the convoy. At night they used the drones to bomb the convoy from the air, first blocking it by destroying the lead vehicles and then destroying more vehicles each night.[239] The Russian forces were unprepared for the asymmetric attack and struggled to respond. Soon, supply shortages, breakdowns, bad weather, and other Ukrainian attacks added to the chaos. Following this excellent demonstration of a porcupine defense, small drone units rapidly proliferated throughout the Ukrainian military. Ukrainian agents and partisans have used them and other precision weapons to hit Russian barracks, ammunition and fuel stores, rail yards, and other targets, and kill military and political leaders in the occupied territories.[240] Small armed drones are so portable that Ukrainian agents used them to destroy bombers and transport planes sitting at Russian military airfields hundreds of kilometers inside Russia.[241] The impact of those asymmetric strikes was as much political as military. They embarrassed the Russian government and fed a perception of vulnerability and dysfunction that undermined the Russian war effort.

The tactics work for hostile forces just as well as for allied ones. American, Israeli, Russian, and other bases in the Middle East have come under attacks from drones and other precision weapons, often from local militias. The low cost makes it easy for the militias to sustain those operations, while their precision means real pain from any attacks that get through. They combine the low cost of IEDs with much greater flexibility, range, and precision. The difficulty of defending against them greatly outweighs the effort in executing them. Insurgents everywhere will use cheap, lethal robotic weapons to burn the ground under their enemy.

In the coming years, it might become next to impossible for conventional forces to hold and occupy territory against determined resistance. A supply of low-cost robotic weapons such as armed drones and loitering munitions could harass an occupying force seemingly from nowhere for years on end. As their cost and size goes down, the difficulty of countering them everywhere goes up. Defense may require massive deployment of sensors and active defenses, which would have to stay a generation or more ahead of the weapons to be effective. Such campaigns could make foreign military bases in some areas unsustainable.

Coercive Counter-Value Strikes

In the pre-dawn hours of September 14, 2019, the sprawling Saudi oil-and-gas town of Abqaiq hummed away quietly, unaware of the devastating attack that was about to strike it. Gated communities of international workers slumbered beneath the desert sky. Only the great towers and domes of the petroleum-processing complex, the largest in the world, seemed awake, spangled with thousands of sodium lamps, a huge illuminated bullseye for the coming drones.

The small attack drones approached in a loose squadron, having flown far across the blackness of the desert, reaching a waypoint miles to the northwest of the facility before turning in toward it on their attack run. By approaching from the northwest, which faced the open desert, they concealed their origin and avoided any air defenses around the facility that might be awake, as those faced the east and south.[242] They picked out their individual targets within the facility and dove to strike home.

From the ground, there was only a moment of noise and a pale streak of reflected light as each diving drone hit like a miniature kamikaze. The initial explosions of the drones' warheads were small, but each hit bloomed into a fireball as the volatile petroleum erupted in secondary explosions. One by one they struck with robotic precision. Three hit giant oil-separation towers, neatly piercing their sides and turning them into gigantic torches. Others hit the rows of hemispherical storage tanks, one drone piercing each unit in just the same spot on its northwest side, throwing sparks into the sky and igniting geysers of flame.

Within minutes the town, the Saudi government, and world markets awakened to a shocking realization. Much of the biggest oil-production facility in the world was offline and burning. The billions of dollars the world's largest arms importer had spent on advanced defenses had failed to stop the attack—or even to provide warning of it. The drones' precision made their attack fiendishly effective for the cost. The world suspected Iran and its Houthi rebel allies in Yemen, but with no returning aircraft to track or enemy commandos to capture, it remained unclear long afterward exactly who was responsible.[243] Iran enjoyed plausible deniability, and it suffered no open war or other direct consequences.

Those kinds of strikes were foreshadowed as far back as the V-1 campaign against civilian targets in Britain during World War II. Modern precision makes them thousands of times more effective, and the Saudi oilfield strikes showed that they are within the power of second-tier militaries. Russia has used precision strikes against civilian infrastructure as a major dimension of its war against Ukraine. They are likely to become a widespread tool of coercion available to more and more players.

Such attacks focus on explicitly economic targets, such as the source of Saudi Arabia's national oil wealth. They are designed for political, not military, effect. They don't strike military forces but rather the things the military is supposed to protect. Because they strike things valuable to the target nation, militaries sometimes call them "counter-value" strikes. They aim to embarrass, create casualties, or inflict economic pain. By striking things that the public is invested in, financially, emotionally, or otherwise, they target enemy will directly. Like cyberattacks, they are tools for coercion, but they offer more certain and dramatic effects. The 9/11 terrorist attacks in New York were a sort of counter-value strike, and their impact was profound.

Rising incidents warn that this trend is ramping up. After repeated drone strikes against Saudi commercial airports and other valuable civilian targets, in 2023 the Houthis made sustained missile and one-way kamikaze drone strikes against international commercial shipping in the Red Sea. Soon after, in January 2024, Pakistan's military made limited strikes inside Iran using, as it said, "killer drones, rockets, loitering munitions and stand-off weapons."[244] As robotic precision weapons become smaller, more lethal, and less expensive, those kinds of coercive strikes will become attractive for more and more state and non-state actors.

Imagine a country locked in a lengthy war with a foreign adversary. The war is unpopular, as the struggle continues for years with no clear battlefield victory. Covert information campaigns have divided the population and sapped public support. The only bright spot for the citizens is that the war seems distant and unimportant in their daily lives. Then, drones emerge from shipping terminals in the country's homeland. Moving unseen at night, they travel to power plants and electrical substations in nearby cities. They detonate at critical places within the machinery as simultaneous cyberattacks amplify their effects. Electrical power services collapse across major metropolitan areas. Responders reach the shipping terminals to find only empty launchers inside abandoned shipping containers. They contain only simple hardware that could have come from anywhere. The drones were widely available models. A previously unknown proxy group claims responsibility. The population becomes outraged. The costs of the war for them have suddenly increased. Protesters fill the streets, demanding an immediate withdrawal from the "forever war."

BY, WITH, AND THROUGH

Proliferation of precision weapons is empowering many new players, including US adversaries. But there is also hopeful news. The United States achieved many of its biggest successes during the War on Terrorism period by partnering with local forces. New robotic systems may be ideal for enabling that successful model.

In 2001 the US swiftly overthrew the Taliban by partnering with the Taliban's opposition, the Northern Alliance. American special operators rode on horseback alongside advancing Northern Alliance fighters. American airpower provided the offensive punch that the Northern Alliance lacked, while the local ground forces were much more effective in waging the political struggle and pacifying captured territory. This "Afghan model" was so successful that it became the basis for the successful campaigns to destroy the ISIS "caliphate" in Iraq and Syria in 2017 and 2018. There, local allies, backed by US special operations forces and airpower, overthrew the ISIS strongholds of Mosul and Raqqa. The US military describes this overall concept as operating "by, with, and through" local allies.[245]

The trick to this model is empowering the local allies without overshadowing them. The strategy leverages their regional standing and political legitimacy, and damaging those can cause the whole effort to fail. That happened in the later years of the Taliban insurgency when the US overshadowed the Afghan government both militarily and diplomatically.[246]

The potentially lighter footprint of robotic fires and maneuver forces may align very well with the "by, with, and through" model of partnering. Loitering munitions and drone arrays can provide similar offensive punch as conventional airpower but with a defter hand than squadrons of strike aircraft. They can be more tailored to smaller operations than conventional air strikes, which have sometimes been compared to sledgehammers striking anthills. Small numbers of US special operations forces, like those that accompanied the Northern Alliance, can operate the robotic assets locally. Large bases and logistics hubs are not required. As a result, they can help empower the local affiliates as they remain in the lead.

Modern war is volatile, uncertain, complex, and ambiguous, and it's getting more so. The future wars in which new robotic weapons will be important will be complicated, messy, and very human. They are unlikely to resemble conventional conflicts of the past. All the belligerents will be empowered by robotic weapons of previously unavailable lethality, intelligence, and elusiveness. Yet they will clash in scenarios as complex as the one facing Lieutenant Santos and her team at the start of this chapter. Future operations must team robotic systems and AI with "strategic corporals" and other human warfighters who can guide them through the complex physical, political, and human maze of twenty-first-century conflict.

The US and its democratic allies have an opportunity to use the inherent advantages of robotic systems to prepare themselves to win those future conflicts. However, we must move forcefully. For the world that results from the currently emerging revolutionary military and technological changes is likely to be more challenging than any we have seen, and time is not on our side.

8

THE TEMPEST OF DISRUPTION

Impacts to Strategy and the Balance of Power

The robotic military revolution's impacts will extend far beyond the battlefield. The social and political earthquakes that came in the wake of earlier technological revolutions in warfare have produced major discontinuities in world history. A new one may be upon us.

An example from history helps illustrate the magnitude of what may be coming. When gunpowder weapons appeared on the battlefield in the late medieval period, European security had been anchored for centuries on the defensive strength of castles. Despite Hollywood's fondness for showing medieval armies storming castles, taking an enemy castle quickly was almost impossible. Wars tended to entail lengthy sieges and limited gains. The Hundred Years' War between France and England, for instance, dragged on for generations as men, their sons, and then grandsons fought in the same war, mostly conducting yearslong sieges. The nobles of the time thought that weapons based on gunpowder, an exotic invention from the East, could help them win the next clash against their neighboring rivals. At first, cannons were siege engines, awkward contraptions that armies had to haul in pieces and construct on location near a castle's walls like a catapult. They were special-purpose tools that complemented more mainstream military systems. Sieges were still long and exhausting, but a cannon's strange appearance and mighty sound did help terrify the defenders into surrendering.

Change came slowly, then quickly. In 1494 Charles VIII of France invaded Italy. It was the first time that an entire campaign was built around cannon artillery, where they provided the main striking power. His generals had matched the new technology with appropriate tactics and all the supporting transport, training, and processes to embrace their full potential. They rolled the cannons and ammunition on two-wheeled wooden carriages, moving quickly from place to place along with a corps of artillerymen to operate them. They rapidly breached the walls of castle after castle. Italian renaissance author Niccolò Machiavelli marveled, "No wall exists, however thick, that artillery cannot destroy in a few days."[247] Charles VIII's army cut through the patchwork of small Italian states like a flying cannonball. Other states rushed to field gunpowder armies. Europe was faced with a military-technical revolution, which one observer described as "like a sudden tempest which turns everything upside down . . . Wars became sudden and violent, conquering and capturing a state in less time than it used to take to occupy a village."[248] Artillery made small states militarily indefensible, leading to the collapse of small kingdoms and their eventual unification into modern nations throughout Europe.[249]

Meanwhile, firearms overthrew the supremacy of knights. For centuries, peasants armed with spears or pitchforks had been no match for heavy cavalry, which were the tanks of the medieval battlefield. But firearms empowered the lowly foot soldier. A commoner could slay a highly trained aristocrat riding a fortune in military technology with the pull of a trigger. The extinction of the knight and empowerment of the commoner helped precipitate the collapse of the feudal system. Thereafter, European history would be shaped less and less by conflicts between nobles and more and more by mass movements and revolutions among the common people. The gunpowder weapons that a few princes had hoped would give them advantage in their battles eventually led to the downfall of their entire political and social order.

In Japan, rulers tried to resist the winds of change. Warlords had used armies with gunpowder weapons to battle for supremacy until, in the early 1500s, one established supremacy over the entire realm. The new military ruler, or shogun, saw that it would be impossible to sustain his position or Japan's feudal traditions while gunpowder weapons were available. The shogunate banned firearms and closed the borders to trade.[250] This froze Japanese society for three hundred years. Yet it eventually led to humiliation. The Western powers continued to advance, and in the 1800s fleets of warships forced Japan to reopen on Western terms. Samurai swords and a few antique firearms were no match for them. The humiliation led to the political collapse of the shogunate and a period of economic and social upheaval as the country struggled to catch up with its adversaries. Japan was at least able to retain its nominal independence—other militarily backward countries fared even worse.

Turning a country's back on change was dangerous then, and it cannot work in the modern world of information technology and global travel. Broad-based technological change, like the industrial revolution and the rise of robotics and AI, cannot be rejected or ignored. But its effects can be anticipated, and societies can prepare themselves for the tempest to come.

The changes in warfare throughout the industrial age, from machine guns and artillery to aviation to nuclear weapons and stealth, have favored the United States and the other leading industrial powers. The US invented most of those innovations and quickly adopted the others. They served to further solidify the US at the top of the global military hierarchy. They made the country more secure. However, the robotic military revolution threatens to be different. This time, we may not be the ones in the driver's seat.

THE FOLLY OF PREDICTIONS

Humans have a terrible record of foreseeing the changes that technology will create. For example, around 1900 many newspapers published depictions of how scientists and futurists thought the twenty-first century might look. The images of future air combat depicted blimps battling with cannons and future soldiers of the air wearing canvas wings shooting at each other with pistols. The visions reflect 1900 more than today. In retrospect the technology looks amusingly antique. But more important was how badly the visionaries underestimated the societal change that comes with advancing technology. No matter what new technologies they tried to imagine, they inserted them into the society with which they were familiar. They assumed a future that looked much like the Belle Epoque, in which gentlemen and ladies strolled the boulevards in genteel elegance and the European empires endured forever. None envisioned that before long those empires would crumble under the cataclysms of two industrialized world wars, followed by the struggles of decolonization. Predictions about the future from later decades look just as naïve: they tell more about the context of the time in which they were made than they do about the world to come.

Robotic warfare will fully come of age in a world that has been transformed by robotics and AI. That world will look unfamiliar, not only physically but politically and socially. The changes will include many things we cannot predict. It is more effective to look at the underlying forces and dynamics that will shape those broader impacts than to make specific predictions. Those suggest that the coming of robotic warfare may upset the order of things even more than industrial-age warfare disrupted that world of 1900.

THE DYNAMICS OF DISRUPTIVE INNOVATION

The robotic revolution in warfare is an example of a disruptive innovation. Harvard Business School professor Clayton Christensen described the dynamics of disruptive innovation in his classic book *The Innovator's Dilemma*.[251] He distinguished disruptive innovations from the more common sustaining innovations. *Sustaining innovations* are incremental innovations that improve a technology or category of products. Military examples of sustaining innovations include more powerful explosives, longer-range missiles, and improved tanks. In contrast, *disruptive innovations* aren't improved versions of anything else. Instead, they introduce entirely new categories of products that never existed before, rendering previous product categories and methods obsolete.

Disruptive innovations usually occur from the bottom up. The early adopters are at the low end of the market. The personal computer, or PC, is a classic example. Before the first PCs, the computer industry was dominated by the mainframe makers such as IBM, Digital Equipment Corporation, and UNIVAC. They built multimillion-dollar computers that took up an entire room and required experts to operate. Their customers were large corporations and government agencies. Only the biggest and wealthiest customers could afford a computer, which gave those customers a competitive advantage. The mainframe makers laid out ambitious visions for the future of computing based on their products. As it happened, the future of computing was indeed soon to arrive, but it would not include them at all.

Small companies like Apple Computer introduced the first personal computers in the late 1970s. Initially called microcomputers, they integrated all the core processing circuitry of a computer onto a single silicon chip called a microprocessor. (Originally used by IBM to refer to its microcomputers, the handy term PC has since become a generic term for any microcomputer.) Those computers with their single-chip processors couldn't compare with the big mainframes in performance, but they cost only a few hundred dollars.

The mainframe companies saw PCs as toys for hobbyists, not serious machines for business. But they enabled vast numbers of new customers to become computer users—customers that were never going to buy mainframes. And the consumer-friendly PCs were powerful enough to meet the needs of those new customers, whose numbers grew and grew as the systems improved. Where computers had previously been only for the elite, the new technology put computers on every desk and in every small business and school. Meanwhile, rapidly advancing microprocessor technology let those computers do more and more of what mainframes could do.

By the time the mainframe makers realized that PCs were an existential threat, it was too late to adapt. Some of them tried to jump on the PC bandwagon, introducing their own microcomputers and positioning them as "programming terminals" for their mainframe-based networks. But customers decided they preferred to have their PCs without the mainframes. Disruption swept the computer industry clean. None of the mainframe companies survived, except for IBM, which no longer makes computers. The computers and mobile devices that we own today are the grandchildren of those early personal computers, not of the mainframes. Today, even supercomputers and AI data centers are essentially vast arrays of PCs.

Disruptive innovations in industry tend to overthrow the previously dominant companies, because those companies find it difficult economically and organizationally to switch to the new low-end products. The big mainframe companies relied on the revenue from selling multimillion-dollar computers to deep-pocketed corporations. At first, PCs were a tiny and risky market not worth their time. The big companies' expensive infrastructure was unsuited to making dirt-cheap microcomputers. They tried to make smaller and less-expensive lines of computers, which they called "minicomputers." But those minicomputers were still based on mainframe technology. They were still big and much more expensive than true PCs.

The disruptive innovation phenomenon revolutionizes industries all the time. Online retail and video streaming are two more recent examples. In addition to mainframe computer companies, we could ask the veterans of department stores, big-box stores, and video rental companies what it's like to be on the receiving end of a disruptive transition. If we can find anyone to ask. Unfortunately, today's leading militaries and their defense-industrial complexes bear worrisome similarities to the computer industry when the first PCs arrived.

The technologies and platforms that provided dominance in the industrial period through the recent past, from the battleship and the tank to the fighter jet and the guided missile destroyer, required size, both in terms of platforms and the industrial systems that made them. If you get the chance to visit an aircraft manufacturing plant or a military shipyard, you will be amazed at their sheer scale. Those facilities are only the tip of a military-industrial iceberg. They require large, specialized parts made in other factories, such as precision casting foundries, and those parts in turn are made of advanced materials like titanium and nickel alloys. And those alloys are made in giant mills that are so extensive that if you stand at one end, you can't see the other without binoculars. Operating all of that requires armies of specialized labor. When an experienced military-acquisition expert looks at a system like a fighter jet or a tank, he or she sees not just a military platform but a whole industrial ecosystem.

All that industrial infrastructure costs billions upon billions of dollars to build and sustain. Defense companies have needed to merge and consolidate to combine their resources. Even the richest nations in Europe cannot afford to build their own military platforms anymore. They combine their resources using multinational consortia. It is very, very expensive to stay on top, and only the biggest defense companies have survived.

Industrial war cemented the biggest industrialized economies as the leading military powers, but that age is ending. Small, precise robotic weapons have created a battlefield ruled by weapon-target asymmetry. Those weapons don't need massive industrial bases. Personal computers broke the computing monopoly of big companies and government agencies by letting people put a computer on every desk and in every home. Robotic warfare is breaking the military monopoly of big powers by letting every state and non-state fighting force afford lethal capabilities that are potent against even the world's leading militaries. The new waves of robotic weapons tend to be much smaller and less expensive, even though the underlying technologies can be more advanced. They substitute precision and intelligence for size and cost, and they leverage commercial electronics and software technologies. In many ways they are the PCs of the military world.

Just like most disruptive innovations, the early adopters are at the low end. It's highly significant that the first war truly built around robotic weapons was conducted by Azerbaijan, a small, emerging state. Recent robotic weapon innovations have included small grenade-dropping drones, FPV kamikaze drones, and satellite-networked attack drone boats. They were introduced by ISIS, an insurgent group, and Ukraine, one of the poorest countries in Europe. Each of them wanted high-end military capabilities but couldn't have them. Early robotic weapons, like the PC, gave them much of the same capability at a much lower price.

Today's dominant defense firms aren't set up to produce small robotic weapons. A loitering munition made by a big defense contractor costs tens or hundreds of thousands of dollars, much like the minicomputers made by mainframe computer companies. Conversely, an FPV attack drone based on off-the-shelf commercial parts costs a few hundred. It might not have all the same performance, but it does the same job—for around 1 percent of the price. And manufacturing small attack drones is well within the capabilities of smaller countries like Ukraine. They contain some high-tech materials and parts, but they require nothing like the industrial base needed to manufacture traditional platforms such as fighter jets. By the middle of 2023, the year after the full-scale Russian invasion, Ukraine had established forty manufacturing sites making small drones. By the end of that year it had two hundred, producing fifty thousand drones per month.[252] Ukraine also started designing and building its own long-range attack drones and cruise missiles, as well as the drone boats that destroyed ships of the Russian Black Sea fleet.

Buying commercial components and performing small-scale production using additive manufacturing, also known as 3-D printing, provides an alternative to capital-intensive, vertically integrated traditional manufacturing. There are only a few parts of small robotic weapons that really require specialized, capital-intensive industry, mainly the microchips. However, those chips are commercially available. Even the most advanced chips used in military machine learning systems are the same ones used for civilian applications like self-driving cars, and they can be readily obtained from an ever-growing number of AI-enabled commercial products.

Major defense companies are laying out ambitious visions for the future of unmanned systems. They envision robotic weapons as accessories for their top-end systems, for instance as less expensive unmanned wingmen controlled by high-end fifth- or sixth-generation fighter planes, a concept that goes back to the assault drones of World War II. Unfortunately, this resembles the way mainframe companies positioned PCs as terminals for their mainframe-centered networks.

However, similar to the early users of PCs, many military customers are likely to decide they prefer their inexpensive but lethal drones without the expensive fighter planes. Larger platforms or motherships mostly provide strategic mobility and sustainment. That is, they let the small and lethal robotic platforms operate farther from home. That is important for the US, with its global security responsibilities and power projection doctrine. But many early adopters may not require intercontinental power projection nor integration with large existing military forces and established procedures. The drones alone may provide more capability, because they can afford many more. Networking them together will multiply their capabilities. By "cutting the cord" to the earlier paradigm, these militaries may seize greater advantage. From humble beginnings, the PC came to dominate the computing world. In the same way, the dominant military systems decades from now may be the far more advanced descendants of small but capable weapons like the armed drones now destroying tanks in Ukraine instead of the traditional military platforms of today.

COLLAPSING THE CASTLE WALLS

The military predominance of the US, as well as other leading powers, is protected by other advantages. But the robotic revolution is threatening to undermine each of those as well.

To begin with, the US military is the world leader in logistics. It mobilized transport and supply on an unprecedented international scale in World War II. Each of its wars thereafter, including Desert Storm and the War on Terrorism, became a showpiece of logistical power. Only the United States can mobilize the vast quantities of sealift, airlift, and aerial refueling that are necessary to support global wars.

But smaller and smarter robotic weapons need less of that. With no need for astronomical amounts of unguided ammunition, fewer large gas-guzzling platforms to fuel, and smaller numbers of humans living in the field, logistics become lighter. Massive logistical enterprises become less important. Instead, they become liabilities. Each of those supply ships, aircraft, and cargo terminals becomes a lucrative soft target for an enemy's precision weapons.

Dominance on the precision battlefield is about sensing and ISR networks as much as it is about precision weapons. The expensive American ISR enterprise, with its spy satellites, surveillance aircraft, and three-letter agencies full of analysts that collect the data and convert it into targeting information, is a big reason the US stays on top. But more and more private space companies let customers anywhere order spy-satellite quality imagery of any region on Earth, delivered directly to their inboxes. Surveillance drones can do much of the work that previously required manned aircraft—and at a lower cost. And robotic weapons increasingly serve as their own ISR platforms. Loitering munitions and small drones that carry high-definition video cameras let users on the ground find and attack targets using the manual "hunt and peck" method. Data networks and AI can vacuum up the data all those systems collect and build a real-time picture of the battlespace at low cost.

Military power also rests on people and institutions. This has been a bigger part of military supremacy than most people realize. Higher education, military academies, training bases, doctrine schoolhouses, and military culture and traditions are all part of it. It's a lot harder to build all of that expertise and discipline than it is to buy some expensive weapons, but large, expensive weapons are not very effective without experts to operate them. However, like the PC, as robotic systems become more user-friendly, they empower nonspecialists. More military training and experience can be captured and embedded in software and AI rather than in expert human brains. It can take years and millions of dollars for militaries to train a new special forces operator or fighter pilot. But they can upload the latest combat AI to a new robotic platform at the press of a button, for almost nothing. Highly trained human warfighters are rare and expensive. They are a handcrafted, luxury product. Combat AI, on the other hand, can be mass-produced. This is another way that robotic weapons, like PCs, will increasingly empower new users and erode the long-standing advantage of the established powers.

Finally, we may hope that access to the most powerful AI models, hosted in giant, expensive data centers, might deliver a new source of military advantage for the US. However, it may be years before centralized AI control proves more helpful than hazardous in many real-world combat applications. As AI technologies advance, powerful capabilities may become less expensive and more accessible. Furthermore, access to such models is already provided to users around the world online, via cloud services. Many rising military powers will be able to buy access to advanced AI capabilities on the commercial market, as they do with satellite imagery today, much more cheaply than building their own.

A FLOOD OF NEW ENTRANTS: RISING POWERS AND NEW GLOBAL ACTORS

As barriers to entry fall, many new types of entities are moving into the market for military power. The new robotic weapons are allowing them to wield capabilities that previously had been the monopoly of the leading militaries. They can increasingly challenge the dominant powers, and the process has only just begun. Here are some examples.

Lower-Tier Militaries

The early part of the first wave of the robotic revolution has featured the rise of new leading weapons producers from mid-tier economies such as Turkey and Iran. Those countries were not previously high-end defense systems manufacturers, but they have become major suppliers in the rapidly expanding global robotic weapons market. The Turkish firm Baykar makes the TB2 reconnaissance-strike drone and has announced an increasing number of more advanced robotic weapons. As of 2023, Baykar signed supply contracts with thirty nations and was building international drone factories in Ukraine and Saudi Arabia.[253] Another Turkish firm, Roketsan, grew alongside Baykar to become a leading producer of small precision munitions, including for armed drones. Iran built a globally prominent robotic weapons industry, initially because smaller asymmetric weapons allowed it to bypass international military sanctions. Its state-owned defense companies, such as Shahed Aviation Industries, emphasized unmanned systems, and many new private suppliers have sprouted within Iran to domestically produce critical components like engines and sensors.[254] In addition to supplying the Iranian military, Iran's producers supply Iranian proxy forces around the Middle East. Despite UN sanctions, Iran assumed the surprising role of a supplier of robotic military hardware to Russian forces during their war in Ukraine. Many more countries and new international companies are setting up production of basic robotic weapons such as small armed drones.

At the same time, more lower-tier militaries have become robotic weapons customers. While the world was focused on more headline-grabbing wars such as in Ukraine, countries around the world flexed their new muscles in drone-powered campaigns. For instance, Ethiopia acquired armed drones from Turkey, Iran, and China. It used them to halt the advance of rebel armies from the Tigray region that were threatening to capture the capital in 2021 and then drive the rebels back into their home province.[255] After compelling an armistice in Tigray, the Ethiopian military launched a new drone war in the Amhara region in 2023.[256] Meanwhile, Morocco acquired dozens of long-endurance drones from Turkey, Israel, and China, and used them to launch an air offensive against the rebels who have long resisted the Moroccan military deep in the deserts of Western Sahara. As the mayor of one rebel city lamented, most of the population fled as everything in the rebel territory became "controlled by drones."[257] The same story is repeating around the world as the affordable weapons deliver extended reach and lethality to nations that never had such capability before.

The process is only beginning. Lower-tier powers are using the new weapons to secure victory in existing conflicts. Before long, they may also flex their capabilities in new disputes with neighboring countries. Like Iran, they may assert themselves against stronger nations they might not have previously challenged. The risks of new interstate conflicts are everywhere.

Insurgencies and Guerrilla Forces

Many of the new customers are national governments seeking novel means to destroy rebel groups. However, in the longer term, insurgencies and guerrilla forces may provide even more natural customers for robotic weapons. The inherent tactical advantages of robotic systems—asymmetric lethality, extended presence and effect, speed of action, elusiveness, persistence, and attritability—are the very qualities valued by insurgent and guerrilla forces. Robotic weapons are fundamentally well suited to the methods of unconventional war. As they become cheaper, small robotic weapons will give those insurgent forces a much more powerful sting than they wielded in the past—while improving their ability to evade detection and avoid decisive battle. Widely varying unconventional forces, such as the ISIS militants and Ukrainian partisans, have demonstrated the power and suitability of small robotic weapons.

When robotic weapons are widely accessible, states may find themselves much more afflicted with uprisings and separatist movements. Greater parity in capabilities will benefit what had been the weaker side. As ideal tools for fourth-generation warfare, insurgencies may use robotic weapons to target critical national assets with counter-value strikes—and to burn the ground beneath national forces that attempt to occupy regions where the population supports the insurgents. Faced with such threats, the national governments may be forced to pursue robotic weapons even more aggressively themselves, seeking to acquire some of the tactical advantages that those weapons offer to counter the newly empowered insurgencies.

Proxy Forces and Militias

Iran has clearly shown how robotic weapons can turn non-state militias into powerful proxy forces. They have given Shia militias like Hezbollah in Lebanon and the Houthis in Yemen the power to threaten the forces of the most modern powers such as Israel, Saudi Arabia, and the United States.

In 2023 and 2024, Houthi forces used Iranian drones and guided missiles to attack military and civilian ships in the Red Sea in support of Hamas, another Iranian client. Iranian-backed militias in Iraq also struck US bases in Iraq and Syria with long-range attack drones and other precision weapons, inflicting casualties and putting political pressure on the American government. The militias provided few targets against which the US could strike back. American and allied satellites and aircraft scoured the Yemeni countryside to find scarce Houthi targets. When the US found a few militia targets to strike in Iraq, the Iraqi government complained bitterly and threatened to revoke the US military's right to operate bases in the country.[258]

Proxy forces, especially non-state ones, are valuable in hybrid or gray-zone conflict and other types of unconventional war. If equipped with robotic weapons, they can produce big effects. They can do the "dirty work" of harassing adversary military forces and executing dramatic counter-value coercive strikes. By claiming responsibility, the proxy elevates its standing and influence while keeping the sponsor nation and its vulnerable targets insulated from retaliation. An ideal proxy force can hit hard but frustrates efforts to hit back in kind, like a swarm of bees. Robotic weapons may make proxy forces widespread players in the age of robotic warfare.

Private Military Contractors and Mercenaries

Once associated with medieval and renaissance warfare, private military companies, or PMCs, have reemerged to play growing roles in modern conflicts. The US military used PMCs like Blackwater to augment its forces during the wars in Iraq and Afghanistan. The Russian Wagner Group led Russia's interventions in Africa and Syria and even spearheaded major combat in Ukraine, such as the assault on Bakhmut in 2022 and 2023. Many more private armed contractors are operating invisibly around the world, as many countries have discovered that they are excellent tools for hybrid and gray-zone warfare.

PMCs are ideal customers for robotic weapons. They offer sophisticated military services, often employing experienced veterans of traditional armed forces and taking on complex and risky missions. Their clients seek to rent specialized skills that their own forces do not possess. Much like the early years of any new technology rollout, there is big demand for contractors who can furnish robotic military technologies along with experienced experts who can employ them effectively. Military advisors and contractors have played key roles in helping militaries employ their new reconnaissance-strike drones quickly in combat in Azerbaijan, Libya, and other countries. Some advisors may have conducted strikes themselves while the host military members learned from them.[259] New robotic weapons will be highly deployable with a low logistical footprint. Soon we may see PMCs offering drone arrays for hire and otherwise bringing a full-service approach to apply robotic warfare to the conflict of a client's choice.

Cartels, Pirates, and Criminal Networks

In 2022, the notoriously brutal Jalisco New Generation cartel attacked a Mexican military convoy using small explosive drones as it approached a town in Michoacán. The explosions killed four and injured six others.[260] A year later, the cartel announced it had activated a standing armed drone unit for strikes against the government and rival cartels. Thereafter, in the first eight months of 2023, the Mexican government recorded 260 attacks by bomb-dropping drones against the Jalisco New Generation's enemies.[261] The ability to strike targets from miles away will open a dangerous new range of possibilities for drug cartels and other criminal organizations, putting capabilities in their hands that previously were the domain of governments. Criminal gangs and cartels, which already threaten to overpower sovereign governments in many parts of the world, may become much more dangerous and powerful. When small and inexpensive robotic weapons allow criminal organizations to threaten military forces, they can easily threaten civilian targets like merchant ships. Piracy, privateering, and commerce raiding could become common practices again, including in service of hybrid or gray-zone strategies.[262]

A WORLD OF DURABLE DISORDER

Robotic weapons are making the world flatter with regard to military power. The so-called "democratization" of military force these weapons are driving will tend to make the world less democratic, more violent, and more unstable.

The primacy of the US and its democratic allies after the Cold War created a "pax Americana" that reduced the worldwide rate of deaths in war to the lowest in at least six hundred years.[263] Even before that, the superpower standoff made the two international blocs the arbiters of war and peace and kept violence in relative check. Advanced states have been the dominant political and military powers. In that orderly world, international law and order has been made and enforced. For the most part, advanced states acted responsibly—or at least predictably. Their advanced status required high levels of education and economic development, and thus they had much to lose. Such states could be deterred from destabilizing the status quo. Mutual deterrence usually prevented direct fighting between the major powers, especially between democracies. Multilateral institutions between states, such as the United Nations, the World Trade Organization, and the International Court of Justice, solidified that stability.

Now the first wave of the robotic revolution is putting unprecedented power in the hands of all kinds of actors who are keen to disrupt that order. They have many different ambitions. And they have access to well-developed playbooks that describe how smaller powers can defeat greater ones. The marketplace of violence may be about to become crowded.

The proliferation of non-state actors with potent attack capabilities will erode the mutual deterrence that limits violence between states. They provide a state many options for sinking a dagger into an adversary while keeping its own hands clean. Many actors may use robotic weapons to "take out" adversary leaders and destroy infrastructure in coercive strikes. When militia proxies, PMCs, and others strike on behalf of a sponsor state, the state can even playact as a supporter of stability. Russia has demonstrated this approach, decrying the violence stirred up by its own proxies such as

Ukrainian separatist groups to justify introducing its own military forces as "peacekeepers." Greater numbers of empowered groups also make it harder to resolve conflicts between states, because more actors can veto an emerging political settlement through violence.[264]

Hybrid and gray-zone conflicts may arise in many regions. Already, the American dualistic concept of being "at war" versus "at peace" is obsolete and widely exploited as a strategic weakness. Some national security strategists predict an age of "durable disorder," when conflicts smolder and spread and can rarely be resolved but merely contained.[265] In that environment, conventional war mutates into many new unconventional forms, and old and new political forces constantly battle. "Shadow wars," where belligerents fight armed conflicts covertly using special operations forces, proxy militias, mercenaries, and other intermediaries, may become endemic. In the same way that military commanders don't want to move forces in the open, political actors may seek more elusive tools because shadow wars are best conducted deniably and covertly. Economic coercion may play a leading role as disruption makes state and personal economic security more uncertain. In all these conflicts, the struggles to control narratives, worldviews, and loyalties will be as important as destroying targets.[266]

The rise of non-state actors serves to erode the "trinitarian" principle that has been the foundation of the law of armed conflict since the 1600s. That principle established separate legal status for the military, the government, and the noncombatant population. It established state militaries as the only legitimate armed forces. Prior to that, it was common for mercenaries, armed religious orders, and other groups to wield military power and conduct their own wars. The proliferation of such groups, and the issues such as religious conflict that animated them, helped make conflicts such as the Thirty Years' War chaotic, uncontrollable, and stained with horrific atrocities. With four and a half to eight million killed, it was the bloodiest European war in history until World War I three centuries later.[267] It was almost impossible to stop and so awful that afterward most non-state armies were abolished on the continent. The "democratization" of military violence powered by accessible robotic weapons could encourage a return to non-trinitarian chaos.

The same trends will empower even small terrorist groups. Security fences, checkpoints, walls, and metal detectors will be less useful. Cheap explosive drones could circumvent such barriers across the world. They are the Molotov cocktails of precision weapons. They may give countless groups animated by hatreds, grievances, grudges, or conspiracy theories the means to murder and destroy from a hiding spot. Any skilled person who can obtain explosives may be able to conduct military-style precision strikes on any target they choose at a range of several miles. That prospect should be sobering for everyone.

FALLING DOMINOES

Throughout the history of robotic warfare, militaries often embraced robotic weapons when they felt they needed to even the odds. The German military fielded its wunderwaffen to help it defeat larger and more heavily equipped European militaries in the Second World War. The US Navy rushed to develop assault drones to quickly build naval strike power following the surprise American defeats at the hands of the Japanese military in December 1941. The militants of ISIS developed armed hobby drones to counter the superior forces of the anti-ISIS coalition. And the Ukrainians looked to robotic weapons to stem the tide of the Russian invasion of their homeland. Robotic weapons provide the asymmetric power needed to counter larger conventional forces. The natural customers for the new generations of disruptive robotic weapons may not be leading, well-equipped, modern militaries but rather those planning to defeat leading, well-equipped, modern militaries.

The United States has been an early leader in military robotics and AI research. Yet, in disruptive innovations, the early pioneers are often not those who ultimately dominate. For example, Kodak, the global leader in film photography, invented digital photography. It failed to adequately embrace it and was buried by the competitors its digital technology helped create.

Because the US and its democratic allies are dominant in today's international order, many of the new players in the age of disruption will be gunning for us. First in line may be some existing players, such as Iran, which has embraced robotic weapons as a key part of its strategy and has already cultivated a network of proxies ready to unleash them. China is explicitly looking to dominate in AI and to counter the US and allied power projection abilities that back up international law and block its expansion in

East Asia. Long-range precision weapons are already the centerpiece of its strategy. Future robotic weapons may be easy for it to adopt. Russia has seen much of its legacy military equipment destroyed in Ukraine. That sounds good for the democratic West, but much like Germany before World War II, it could free Russia to rebuild its military on a new model. It is massively increasing production of robotic weapons due to the war and could end up leapfrogging NATO in embracing robotic warfare.

Traditional US adversaries such as hostile authoritarian states might soon themselves be overtaken. They are more rigidly hierarchical than democracies, less capable of change, and more reliant on internal repression for control. As the robotic revolution progresses, they are even more susceptible to newly empowered liberation movements and insurgencies. Some may be consumed by internal conflicts and may lose control of their agenda to non-state powers.

Many rising states and non-state forces may target the US and its allies as well, because we stand in the way of their disrupting the international order. Non-state forces and emerging powers don't need to achieve military supremacy to dictate the future of robotic warfare or to apply coercion to the leading states. They can seize control of the agenda with lethal attacks that force the leading states into reactive mode. Al-Qaeda transformed the agenda of the leading powers in one day in 2001, and it pushed the US and its allies into a two-decade War on Terrorism that none of them had contemplated a few days before. Robotic weapons could enable new powers to impose their will, even from a small power base.

WE ARE IN THE BULLSEYE

Fourth-generation warfare reminds us that war is politics, and fighting an enemy's military is not always central to winning wars. The airpower theory that drives the US Air Force and other leading air forces recognizes the same thing. Strategic airpower theory is founded on the recognition that air attack offers the chance to bypass an adversary's military forces entirely to strike directly at the vital functions that enable its war effort. The functions are usually conceived as a set of nested "strategic rings" of increasing importance, in the shape of a target. The outermost ring consists of the military forces. Middle rings include production infrastructure and important systems such as command and communications networks. The innermost and most vital ring, depending on the specific model, is either the national leadership or the will of the people. Modern democratic states observe the trinitarian principle and refrain from directly attacking national leaders or populations, stopping at vital systems and infrastructure. However, fourth-generation warfare urges adversaries to target the innermost ring through information, media, political subterfuge, and intimidation. Robotic weapons are increasingly enabling adversaries to target it with physical attack.

Many civilians and political leaders in the developed world are used to thinking of wars as unpleasant things that happen far away. Except for those who have friends or family serving in the military, we are accustomed to being bystanders. However, politicians and the public may increasingly be the focus of hostile operations, and robotic weapons will put us in range of many hostile actors. Non-state forces can bypass state militaries to strike economic targets and threaten citizens, and conflicts between states could come to resemble mutual civil wars. Civilian populations will be a center of gravity for many conflicts. The days of security when citizens of the leading democratic states may choose not to be involved in military affairs may end. In short, the US and its allies have a big target on our backs, and it encompasses us all. Even if we as private citizens are not interested in robotic warfare, robotic warfare is interested in us.

History shows that we cannot predict in detail the long-term outcomes that the robotic military revolution will have for society and geopolitics. But its underlying dynamics suggest that it could be more inherently disruptive to established powers than any other military-technical revolution in centuries. US security and global stability are protected by a walled fortress of military advantage. Its walls are built on factors such as industrial strength, superior access to information, military expertise, and mastery of industrial-age logistics. Robotics and AI are empowering many adversaries with weapons that could breach those walls of advantage and unleash a tempest of disruption that could turn our world upside down.

DRONE RAID: A POSSIBLE NEAR FUTURE

Sirens wail across the wide-open tarmacs of the base. Commander Cheng races along the side of one of the aircraft hangars and takes shelter, sweat tricking down his back. He shouts orders to the aircrews and maintenance technicians, some of whom take shelter inside the hangars while others run past, carrying anti-drone jamming guns. The angry whir of drone motors echoes from all around. The frenzy of alarms and explosions seems surreal against the backdrop of a line of palm trees and the placid expanse of the bay just beyond, becalmed beneath the baking afternoon sun.

The attacking TTM forces caught the base by surprise. They overwhelmed the meager perimeter defenses and precisely hit the base headquarters, the security forces post, two electrical substations, the aviation fuel depot, and a half dozen other targets within minutes. Not far away, Cheng sees the remains of a parked trainer aircraft burning under the open sky. Between the hangars and the shoreline are three dark-green short-range air defense vehicles, meant to deter attacks like this. Their squat, eight-wheeled shapes are covered with radars and weapons. Every few moments, one of the computer-controlled 30-millimeter guns on the vehicles shakes the air with a blam-blam-blam and disintegrates a drone somewhere overhead. Then one of the laser weapons fries a drone in the air like a mosquito touching a bug zapper.

Many of Cheng's personnel have never experienced such a large-scale attack, but he has. On his previous assignment, he helped evacuate friendly personnel and equipment from beleaguered Malaysian naval bases. Militant drone attacks had been unrelenting. The Chinese regions that split apart in the wake of the financial crisis all had hyperscale drone factories that were producing weapons to use against each other. Many of those weapons were also ending up in the hands of rebel movements across East Asia. That had destabilized the whole region. With the closure of commercial shipping lanes from China, consumers from Malaysia to the US could no longer find half the everyday items they once took for granted, but drones and other weapons were never in short supply. Before the trouble started, Cheng had vacationed with his family on the beaches of Phuket, in Thailand. Now much of Southeast Asia is too riven with conflicts to allow leisure travel. Still, he'd expected his new assignment would be safe.

Small shapes emerge from around the corners of nearby hangars and zip toward the armored vehicles. The military had bought the vehicles when it still thought of the drone threat as an air defense problem, the "air" being overhead. As a result, their weapons could not depress below the horizon. The FPV drones race across the tarmac at kneecap height and strike the vehicles in their wheel wells. The moment it takes the small drones to travel from the edge of the hangar to the vehicles is not enough time for the defenders to bring handheld jammers to bear. Explosions throw smoke and debris. Cheng and his personnel dive for cover.

With the defense vehicles disabled, Cheng yells for his people to take shelter inside the nearest hangar. The smoke offers a moment of concealment as they run. The handheld jamming guns might be of more use in the confined space inside. If only they had been better prepared! The naval training base has no combat units and has never been heavily defended. Service members like Cheng know it largely for its beautiful weather and seaside campgrounds.

Only a year and a half ago, six months after his arrival on station, he first heard of the militia when they bombed police stations in the backcountry. Nobody could have guessed then what political cause they supported. At first Cheng had assumed they were connected to the cartels, who must have finally gone crazy enough to strike inside the country. When the TTM proclaimed itself after a string of bigger attacks, his reaction had been to scoff that their name was silly, because there had never been any tigers in Texas. Yet eighteen months later, no one is laughing at the Texas Tigers Militia. Could this really be happening?

As he helps the other US Navy men and women in his unit push the hangar door shut, Cheng catches a surreal glimpse of the downtown skyline and the harbor bridge behind the burning vehicles and the palm trees, across Corpus Christi Bay from the naval air station. It's like looking at the past through the chaos of a present that seems to be falling apart.

9

BELUA IN MACHINA

Machines, Morality, and the Heart of Darkness

On the morning of March 16, 2022, middle-aged men Vitaliy Kontarov, Ihor Moroz, and Grigoriy Golovniov were each foraging for supplies outside the Donetsk Regional Drama Theatre in Mariupol, in southern Ukraine.[268] The stately nineteenth-century theater had been turned into a refugee shelter as Russian forces closed in on the city from east and west. Hundreds of other civilians, mostly women and children, huddled inside, including survivors of a maternity hospital hit by Russian bombs a week before. The theater's set designer had painted the word "Дети," or "children," on the plazas on two sides of the theater, in letters large enough to be visible in satellite images.[269]

The men heard the growl of a military jet passing overhead, which was familiar because the Russian air force was bombing the city constantly. They told what happened next to human rights investigators weeks later. "We heard planes . . . I saw two missiles fire from one plane towards the theatre," said Vitaliy. Ihor recalled, "It all happened in front of our eyes. We were 200 or 300 meters away [when] the explosion happened . . . I could hear a plane and the sound of bombs dropping. Then we saw the roof [of the theatre] rise up." Grigoriy also reported the plane, and the detonation: "I saw the roof of the building explode . . . It jumped 20 meters and then collapsed." Survivors from deep inside the theater reported bring thrown by explosions as the top floors of the building were ripped away along with the families sheltering there. Survivors from the basement emerged past hellish scenes of wounded amid the rubble. One man emerged to find people bleeding, some with open fracture wounds. "One mother was trying to find her kids under the rubble," he recalled. "A five-year-old kid was screaming: 'I don't want to die.' It was heartbreaking."[270] The government of the besieged city estimated that three hundred had been killed.

Analyses of the blast pattern and eyewitness testimony suggested that Russian 500-pound (250 kg) laser-guided bombs had detonated just inside the roof.[271] There was no hint of Ukrainian military activity in the area and no reason a drama theater would be on a list of strategic targets. The use of laser-guided bombs meant the pilots would have had a clear view of the target before they struck.

There would be no further investigation. Soon after the strike, advancing Russian forces completed their encirclement and fought their way toward the city center. Within two weeks they occupied the waterfront district where the ruins of the theater stood. Ninety percent of the city's buildings were damaged or destroyed. Surviving eyewitnesses escaped the siege by fleeing for days through occupied territory. After consolidating their grip on the ruined city, Russian occupation authorities cleared the site and prepared to rebuild the theater as a showplace for Russian culture.

The early inventors' vision for robotic weapons was to introduce precision and certainty to the chaos of war. Armed with those products of superior science and reason, their theory went, the more technologically advanced and civilized countries could defeat the more aggressive ones. By making war ever more a contest of machines they would sanitize it, lessen its savage aspect, and hopefully pave the way to eliminating war as a human activity.

It was a good idea, and it could still work. Unfortunately, two complications arose. The first was that technological advancement turned out not to be incompatible with aggression and savagery. Nazi Germany made that painfully obvious, and other examples followed. The second is that the disruptive nature of robotic weapons is putting them into many new hands, and many of the new users do not care about certainty and sanitization. If they are controlled by those who don't share the rationalist, scientific approach to warfare, robotic weapons and AI could empower the barbarism that they were intended to defeat. We should worry about how we will employ robotic weapons. We should worry a lot more about how others who are very different may employ them, including against us, and what that will mean for the future of human conflict.

TERROR AS TACTIC AND STRATEGY

Militaries have long employed wanton savagery and terror as tools of war. Long before US airpower theorists determined that an industrialized enemy could be subdued scientifically by precisely destroying key industrial and logistical choke points, armies understood that enemies could be subdued through fear. In ancient times it was common for aggressors to threaten massacre on towns or cities that did not surrender. For example, Genghis Khan punished cities that attempted to resist his armies by burning them to the ground, slaughtering all the survivors, and building pyramids of human heads as warnings to others. Despite Mongolia's small population, his armies were able to quickly subdue much larger empires, in part through intimidation and the sheer terror that went before them.

When the German armies marched through Belgium on their way to France at the start of World War I, the German command ordered them to make Belgium surrender quickly through a campaign of terror. The military answered resistance with merciless atrocities, including the destruction of the town of Leuven by rampaging troops. As one German officer said, "We shall wipe it out . . . Not one stone will stand upon another. We will teach them to respect Germany. For generations people will come here and see what we have done."[272] The Allies called this policy *schrecklichkeit*, which is the German term for horror or frightfulness. In World War II the Nazi regime applied this policy far more widely. It sometimes used its robotic wunderwaffen in support of terror campaigns. The V-1 campaign to destroy London was an act of vengeance meant to horrify Britain into surrender. The robotic ground vehicles were tools of terror in crushing the Warsaw Uprising.

In Ukraine, starting in February 2022, Russia made strikes against civilians on a massive scale, including with its precision weapons. It poured barrages of rocket artillery into civilian residential districts.[273] It used guided missiles to strike hospitals, shopping centers, apartment towers, restaurants, and many other civilian targets. Russian precision strikes against utilities and power stations during the winter also attempted to apply civilian suffering to break Ukrainian will.[274] Analysts noted that the Russian military expended a high proportion of its cruise missiles and other precision weapons against Ukrainian civilian targets.[275] That was consistent with other recent Russian practice. As a new prime minister in 1999, Vladimir Putin established his reputation by brutally suppressing the Chechen rebellion, using mass artillery, ballistic missiles, and TOS-1 thermobaric weapons on the capital city of Grozny. Afterward, the United Nations described Grozny as the most destroyed city on Earth. In Syria, Russian air strikes targeted markets and hospitals in rebel-held areas, helping Bashar al-Assad's government to win not through battlefield victory but by making life so horrific its enemies would surrender to make the suffering stop.[276]

Terror bombing has a generally poor record of producing military victory, but it is nonetheless widely used. It requires little expertise, so poorly trained militaries can carry it out. Its conspicuous violence gratifies authoritarian leaders and their domestic supporters.

Many other states have recently used military force in service of ethnic cleansing and other campaigns of terror targeting civilian populations. Campaigns in the former Yugoslavia, in Sudan, East Timor, Myanmar, and elsewhere exhibited the strategy. The war of the Ethiopian government against Tigray may have killed many times more civilians than the war in Ukraine. Human rights observers accused the Ethiopian government of making widespread drone strikes against civilians in Amhara.[277] The Moroccan drone campaign has similarly been accused of targeting noncombatants, to drive the civilian population of Western Sahara off the land.[278]

States have also used military force against their own populations outside of war to subjugate and control them. The Tiananmen Square massacre, in which the Chinese government used army units including tanks to crush civilian democracy protests in central Beijing in 1989, was a graphic example. Extreme cases, like the bloody reign of the Khmer Rouge in Cambodia in the 1970s that killed almost a quarter of the population, are just the tip of the iceberg. Other cases of authoritarian states using military force to intimidate and cow their own people are too many to list.

When the perpetrators are non-state actors, the use of terror and savagery may be even worse. Such groups are even less bound by the expectations placed on state militaries and more empowered to unleash atrocities against populations the group finds objectionable. Modern fighters in groups such as ISIS and Boko Haram have shown no hesitation in massacring civilians and committing atrocities as core parts of their strategy. Brutal non-state violence is not a new phenomenon. It was a Catholic monk leading volunteer forces in a religious war, the Albigensian Crusade, in France in the 1200s, who first uttered the command, "Kill them all and let God sort them out" outside the rebel town of Béziers. His troops massacred up to twenty thousand townspeople.

In short, human control over robotic weapons does nothing to ensure ethical use if those humans are trying to target noncombatants and commit atrocities. Brute force has always been more readily available than military skill. Indiscriminate suffering is easier to apply than highly targeted precision effects. As robotic weapons proliferate to new users, many of them will be comfortable using them as terror weapons. For such users, robotics and AI could provide new opportunities for terror and mass destruction.

ROBOTS AS FUTURE TERROR WEAPONS

Some analysts fret that robotics and AI could produce more horrifying weapons. However, a precise and selective weapon, like a samurai sword, isn't horrifying in and of itself. But when precision and selectivity are not the goal, robotics and AI may provide several new dimensions that malevolent users could exploit to maximize horror.

Design for Intimidation and "Robot Fright"

Robot designers use anthropomorphism and other cues to shape our reactions to robots. Today they employ familiar shapes, pleasing human voices, soft colors and forms, and other cues to help them appear safe, even charming. Humanoid robots are often portrayed with a pleasingly human faceplate that serves no purpose other than aesthetics. Those design choices help comfort humans with whom they interact, decreasing anxiety and promoting trust. That makes sense for harmless service robots or for robots that serve in settings such as airport security checkpoints, where they may interact with families and children.

However, designers could easily take that principle in the opposite direction, using design to increase intimidation and amplify feelings of danger and fear. Soldiers who encountered early tanks often spoke of "tank fright" that paralyzed them with fear due to the tanks' unfamiliarity and menacing appearance. Designers could seek to amplify "robot fright," especially for terrestrial robots or drones meant to be visible to enemy soldiers or civilians. They could emit frightening sounds. Arbitrary faceplates could be intentionally horrifying or inhuman. Robots could use visible cues that communicate danger, such as aposematism, the sharp edges and high-contrast colors that subconsciously warn of danger in venomous insects and reptiles.

Software Mutability

Predictability is a core element of trust. That is especially true with robots. However, unlike living things, a robot's nature is not an immutable part of its being. A robot's operating software can be replaced at any time, changing everything about its behavior and rendering all past experiences with it irrelevant.

Once we become familiar with a new type of living thing or inanimate object, we can predict its behavior and understand how to deal with it, because its nature is immutable. An animal such as a dog or a shark behaves according to its nature. A dog might be dangerous because it can bite, but we can manage that risk because we know that a dog will likely growl or bark to communicate aggression, or it will wag its tail to signal friendliness, and so on. The same predictability allows us to handle a potentially dangerous inanimate thing like a hot stove or a handgun.

A robot's software can change without warning or external indication. The change could result from a faulty update or a cyber intrusion. Due to software mutability, any robot we know one day may not be the same the next. A benign robot could turn into a killer at any time.

Scientists and philosophers have debated to what degree the human mind is part of our physical being, specifically our brains. Philosopher Gilbert Ryle labeled the idea of the mind as a separate thing as "the ghost in the machine." But for robots there is no debate. A robot's hardware and software are definitely separate, and its "mind" is easily changed.

Arthur Koestler, in his 1967 book *The Ghost in the Machine*, observed that the human brain contains earlier and more primitive structures in common with lower animals as well as more advanced ones that control rational thought. The primitive layers can sometimes overpower rational logic. Hate, anger, and fear can overcome reason under some circumstances. In robots, violence can replace reason at high speed. In one simple example, Israeli security services planted explosives in thousands of pagers and walkie-talkies issued to Hezbollah militants, turning harmless electronics into deadly weapons. The explosions in September 2024 killed more than thirty militants and injured over three thousand. Software changes could transform more sophisticated robotic equipment into hazards or remote-controlled weapons. Software mutability may limit human trust in robots for all time. But malevolent human controllers could manipulate it purposely to confuse and terrify in new ways.

Mad Dog Attacks

Making a good robotic weapon capable of high selectivity, like making a good anti-cancer treatment, is hard. Making a bad one that is indiscriminate is easy, and bad ones are likely to appear first. A small autonomous explosive quadcopter, the Turkish Kargu-2, may have been used in the Libyan Civil War in 2020 to make an unsupervised attack, with uncertain results.[279] More examples are sure to come.

An armed robot that is deadly but indiscriminate would be considered unreliable and ineffective in Western models of precision warfare. However, it could be well suited to terror missions directed by unethical states or non-state groups. For them, deadly and indiscriminate is the goal. An armed robot or a group of armed robots could be set loose as an out-of-control hazard, akin to releasing packs of rabid dogs to chase and maul pedestrians.

Swarms as WMD

Under military control, semiautonomous munition swarms may be powerful and practical both in overwhelming individual targets with swarming by fire attacks and with striking widely dispersed but distinct targets such as mines in a minefield. However, using them for less structured missions and in more ambiguous environments will be challenging both technically and tactically for the foreseeable future. Unsupervised AI cannot yet manage the complexities of the battlefield, and the unpredictable emergent behavior of self-organizing swarms will add even more chaos. Such swarms could easily be just overempowered groups of bad signature-seeking weapons. The difficulty in wielding autonomous munition swarms means that many of them may be hazards more than effective weapons on the battlefield, akin to clouds of poison gas.

However, their combination of lethality, unpredictability, and poor selectivity could make them powerful weapons of terror against civilian targets. Their military shortcomings would be not bugs but features. Like other weapons of mass destruction, they may be more useful to terrorists than to professional militaries. Very crude target signatures like "all humans" or "anything that moves" could make them appealing tools for terrorists or for authoritarian forces that use illegal indiscriminate mass-casualty attacks as an instrument of policy.

DETACHMENT AND DEPERSONALIZATION

Many fears about unethical use of robotic weapons have centered on the idea that military robotics will promote detachment and dehumanization among those who use them and thereby encourage war. Here again, the concern is valid but likely misdirected.

Western militaries are well past the peak era of impersonal "push-button" war. The depersonalization of war reached its high point in the strategic bombing of World War II and in the nuclear drills of the Cold War. In World War II, the crews of strategic bombers had very little interaction with the ground. They often dropped their bombs on map coordinates, or on a radio beacon that had been dropped in advance by a pathfinder aircraft. Their unguided bombs wreaked indiscriminate destruction that they never witnessed firsthand. The flak gunners in turn filled the sky with shrapnel but rarely saw the individual bombers they shot down. Poet and Air Corps veteran Randall Jarrell described the alienation of that kind of battle after the war:

In bombers named for girls, we burned
The cities we had learned about in school—
Till our lives wore out; our bodies lay among
The people we had killed and never seen.[280]

There is no act so isolated from its consequences as launching nuclear ballistic missiles. On a Cold War ballistic missile submarine, for instance, the missile launch room was perhaps the most comfortable on board. In training, an officer sat in the quiet, air-conditioned space on a soft chair, surrounded by computers, and squeezed a trigger every time the control panel requested "intent to fire." The presence of other officers, the pressure of fail-safe procedures, and the tension of the moment didn't remove the surreal disconnect between the movement of a finger and the destruction of military bases or cities thousands of miles away. Fortunately, that scenario never became reality, in part because of the responsibility of the leading nuclear powers.

Today, war is much more personal, as the example of the "strategic corporal" covered in chapter seven illustrates. Contrary to some expectations, robotic weapons have made war more personal than ever before. Some critics suspected that employing reconnaissance-strike drones in the War on Terrorism would produce a detached or "PlayStation" mentality. However, the reality was just the opposite. Drone operators were exposed to close-up scenes of graphic violence more often than even special operations forces on the ground. The need to carefully select the right individuals, and then choose the moment to strike with the least collateral damage, meant that the drone operators had to watch their potential targets doing ordinary things such as having breakfast or playing with their kids, creating a sense of intimacy unknown to prior warriors. The consequences of watching such scenes, and then witnessing the aftermath of their own split-second decisions when the time came, was often tortuous. For example, one operator conducted a strike that killed a terrorist network member while sparing his child. With horror, he watched as "the child walked back to the pieces of his father and began to place the pieces back into human shape."[281] Reconnaissance-strike pilots and weapons operators reported some of the highest rates of emotional distress, largely because of the surreal intimacy their technology provides and that their precision doctrine demands.

The reconnaissance-strike example illustrates that war remains awful but also the extent to which today's democratic militaries will push their warfighters and technology to achieve precision and minimize injury to noncombatants. It took almost ten years for intelligence and targeting capabilities to catch up with the precision of the weapon system, but by 2012 civilian collateral deaths had dropped almost to zero, with strikes achieving "near-unerring accuracy," according to a Brookings Institute study.[282] Ultra-low collateral-damage weapons and other advancements are continuing to improve it.

The pilots of FPV drones in Ukraine got an even more close-up view of the enemy, often seeing enemy soldiers' last moments of terror as the drones chased them down in the open or found them inside trenches or hiding places. The pilots reported that they felt less of the emotional trauma or moral injury reported by US drone pilots in the War on Terrorism because they were targeting enemy soldiers invading their homeland.[283]

Critics of robotic weapons say, for instance, that "autonomous weapons would lack the human judgment necessary to evaluate the proportionality of an attack, distinguish civilian from combatant, and abide by other core principles of the laws of war."[284] The US and other democratic militaries agree. The increasing complexity of modern war argues against the utility of hands-off, push-button warfare. Robotic military operations in the volatile, uncertain, complex, and ambiguous environment of modern war require human oversight. While AI may help reduce burden on operators, the trend is for the role of the "strategic corporal" to continue in the robotic age.

PRECISION ENABLES ETHICAL USE

Improved ethical conduct under the laws of war—that part of international convention that regulates the use of armed force—requires precision. While they do not consist of a formal set of codified laws, they enshrine three principles for the legal use of weapons in war. First is *distinction*, meaning that strikes must target military objectives and not civilians or civilian objects. Second is *discrimination*, which means that weapons or strikes must address discrete military targets and not larger areas in which military objectives are combined with civilian populations and civilian objects. Third is *proportionality* or restraint, which means taking precautions to minimize expected civilian loss of life and avoid excessive injury or damage to civilian objects.

Practically speaking, those are all statements about precision. The precision and certainty that give weapons greater military effectiveness are the same qualities that enable more ethical use. Therefore, the goals of robotic weapons development and those of military ethics are closely aligned. For instance, some have asked why the US didn't bomb the rail lines leading to Auschwitz during World War II. Unfortunately, we had no means to strike them precisely then. The best that long-range bombers might have done was carpet-bomb the entire area, including the surrounding towns. If any bombs were lucky enough to damage the tracks, they would have been repaired within hours. Today we could destroy them with precision weapons without killing hundreds of civilians nearby. Such better ethical options are made possible by the improved precision, or more comprehensively, the improved selectivity, enabled by robotic weapons.

However, those improved options are only of interest to militaries who care about ethics. Detached, push-button weapons could appeal to those conducting terror campaigns and atrocities for whom selectivity is unimportant and dehumanization is a helpful feature. If they serve ends defined by a murderous regime, terrorist group, or criminal cartel, issues such as whether these weapons are controlled via human *in* the loop or human *on* the loop are moot. The greater concern, therefore, is not that robotic weapons may depersonalize the means of combat. It is how we define the ends for which robotic weapons are used, and who will get to choose those ends.

DEFINING ENDS AND THE PRINCIPLE OF MISSION COMMAND

In the most complex and ambiguous environments such as combat, robots may always need some human teaming. However, as technology improves, we will see robots functioning on their own in a wider range of tasks, in everyday life as well as in the military. It will be increasingly apparent that whether they act ethically, and otherwise perform as we intend, has less to do with whether a human is supervising everything robots do as they do it and more with how we define the ends they pursue and the rules they follow to get there.

Military leaders have had to address that same question regarding the human warfighters under their command, as changes in modern war have required subordinates to act with ever more autonomy and initiative. As a result, professional militaries have developed and refined the principle of mission command.

Mission command is a core tenet of modern military leadership. It focuses on communicating the commander's intent, providing an understanding of the purpose or desired ends of a mission rather than specifying in detail the means to accomplish the necessary tasks. That flexibility enables subordinates to use their skills, disciplined initiative, judgment, and creativity to determine how best to accomplish the desired ends in the face of changing and unpredictable circumstances.[285] Commanders define the desired end state, directing "what" and "why," while subordinate leaders devise "how."[286] As General George Patton put it, "Never tell people how to do things. Tell them what to do and they will surprise you with their ingenuity."

Mission command implies trust, and therefore it is much more widely embraced by the militaries of democratic countries than by autocratic governments that distrust their people, including their soldiers. Established doctrine provides specifics on how desired end states should be defined, how risk should be assessed and assigned, and many other details of mission command. Militaries can adapt this approach for the lower levels of initiative, judgment, and creativity that AI can practically have and the more finite empowerment of robotic weapons.

As a simple example, suppose that aircraft are ordered to bomb an enemy aircraft carrier. However, the pilots are unable to find the carrier because of cloud cover or some other issue. If they see a different enemy carrier nearby, they could attack it instead. They might also attack other enemy naval ships if they find them, as secondary targets. Those would all be consistent with the commander's intent, which was, technically, to attack the enemy fleet with targets in order of priority. Mission command increases certainty and effectiveness because it allows the warfighters closest to the action to adjust their plan of attack autonomously to best attain the desired outcome. It also enables military ethics. For instance, if what was expected to be an enemy aircraft carrier turned out on closer inspection to be a civilian passenger liner, the pilots could abort the attack consistent with the commander's intent.

The clear definitions of authority that are specified by mission command can apply to AI-enabled robotic weapons. If the weapon has the required capabilities, the flexibility to exercise such authority within tightly constrained limits can produce greater certainty and control of outcomes. For instance, if a robotic weapon's only task is to hit a set of geographic coordinates, that involves very limited authority. It will hit that spot whether the intended target is there or not. However, if carefully empowered, it could look for the intended target in the vicinity of those coordinates and modify the attack based on circumstances to produce the intended result and avoid unintended consequences.

Mission command recognizes that human subordinates have their own agency. A soldier is not a weapon wielded by the commander or by the state but a moral agent and a combatant under the laws of war. A soldier is a sentient being and ultimately decides what to do, including whether to obey orders.

Mission command authority will necessarily be more constrained for robots and AI because they are weapons, not independent moral agents. There is no applied science of sentience that could enable the design of robots able to exercise agency, especially in a combat environment. Thus, there is no justification for expecting any current or envisioned military AI to choose its own ends. That would position AIs as artificial combatants. They lack the ability to understand the nature and consequences of their actions. Therefore, in contrast to employing them as very smart tools to precisely enact human will, expecting them to exercise their own agency would be akin to arming the legally insane. It would lead to losing control and certainty, thereby departing from the purpose of robotic weapons. It would be an egregious case of overempowerment and itself an unethical decision. Properly adapting the doctrine of mission command can help prevent this.

LAYING A FOUNDATION FOR ETHICS IN ROBOTIC WARFARE

We can combine the framework of mission command with two other complementary frameworks introduced in chapter two to yield a stable foundation for responsible autonomy for AI-enabled weapons. Mission command provides a framework for applying limited autonomy to execute a mission whose ends are defined by an appropriate human commander. The framework provided by the military targeting process adds the considerations relating to the selection of targets and the application of force. It similarly assigns ethical responsibility to the commander and notes that delegation, whether to a human or an AI, does not absolve the commander of that responsibility.

Lastly, the framework of medical ethics introduced in chapter two addresses the case of ethical responsibility for situations where every individual act by a nonhuman agent, such as an anti-cancer drug, cannot be individually overseen and approved by the responsible human. It focuses on the principle of selectivity, and it can be directly adapted to robotic weapons. The combination of the three frameworks of military targeting, mission command, and medical ethics, with their extensive supporting literature, provides the structure for addressing moral questions about military robotics in a more rigorous way than often happens in public debate. They define guardrails for autonomous decisions and clear assignment of responsibilities that can avoid confusion about the ethical accountability for robot actions.

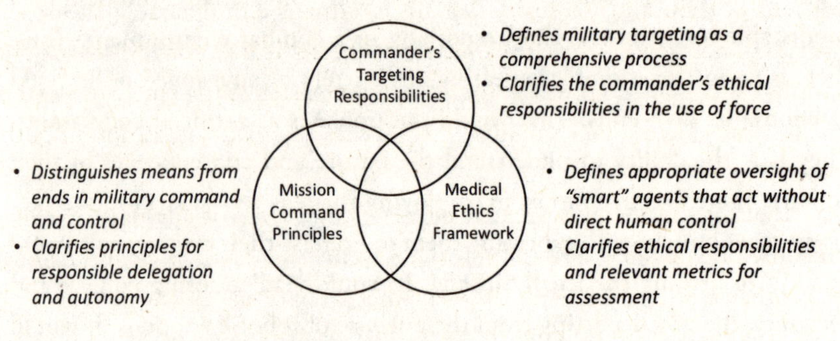

The intersection of three established frameworks that provide a foundation for the ethics of AI-enabled robotic weapons.

All three frameworks share the conclusion that ethical responsibility cannot be delegated to something that lacks agency. Since true agency is only a hypothetical possibility for AI, human commanders and operators will retain ethical responsibility for the selection and employment of robotic weapons for the foreseeable future. The frameworks provide the details of how to do that in practice. If we apply them carefully, we will use the improving capabilities of robotics and AI to enact an ethical commander's or operator's intent with greater certainty and precision and thereby enable more ethical and lawful military action.

For the near term, we will need to define intent narrowly and explicitly in ways that are technically digestible enough for current AI. For instance, "Destroy the enemy tanks inside this defined boundary." As AI improves and robotic systems become more autonomous, we may expect them to interpret human intent more broadly, in ways that more closely resemble the ways that military leaders delegate to human subordinates. That may require us to increasingly confront a fundamental gulf of understanding between human beings and machines. As we try to communicate our deeper purposes to AI, it will require us to grapple with the factors that motivate the human impulse to war and violence and the dangers of asking machines to interpret them.

ROBOTS AND AI ARE FUNDAMENTALLY ALIEN

Homo sapiens have been the only form of higher intelligence on Earth for tens of thousands of years, ever since humans eliminated all the other species in our genus. Thus, the only form of intelligence with which we are familiar is our own. However, for all human existence to come, we may be living alongside digital computer intelligences of our own creation in both peace and war.

That may make us aware that our particular form of intelligence is not the only one possible, and that it might even be unusual. We can imagine that other intelligences with different origins and natures might exist on distant planets. Yet no living thing may be as alien to us as digital AI. All living things we know share some kinship. Even the sea creatures that glow in the permanently dark abyss of the ocean, or the bacteria that make colorful blooms in near-boiling hot springs, use biology and organic chemistry akin to our own. Computer-based AI, in contrast, is different down to the molecular level. It is based on silicon crystals rather than on carbon-based organic molecules. Its fundamental logic is electronic and digital rather than biochemical and analog. It can be turned off and on again. Its software is separable from its hardware and can be changed at any time. It has almost nothing in common with us as living creatures. It is not even alive at all. Digital artificial intelligence is more fundamentally alien to us than any form of life that might exist in the farthest reaches of the universe.

The Polish author Stanislaw Lem is known for works that explore the concept of alienness. His erudite 1968 novel, *His Master's Voice*, explores the potential difficulty humans might have in understanding a signal from an extraterrestrial intelligence. In it, he notes that the simple telegram "grandmother dead funeral Wednesday" can be understood in any human language because it consists of universal concepts familiar to all human life: motherhood, death and its ritualization, and the passage of days.[287] But Lem pointed out that organisms exist on Earth that do not experience one of those concepts, because they live in eternal darkness, do not reproduce sexually, or are biologically immortal like cancer cells. He suggests that extraterrestrial intelligence might exist that is so utterly alien that it does not experience any of them. What common concepts could then allow us to understand it, or it to understand us? Yet computer-based digital AI is such an intelligence.

Deep-learning AI behaves in ways that mimic humans because it has been trained to mimic humans. However, deep learning itself is not a human process. Deep-learning AI systems can identify and replicate subtle patterns from human-generated training data, but that does not make them humanlike. Robots do not spontaneously exhibit human behavior. Engineers can make a robot in a human shape, and make it move as if it were dancing, but it is not really dancing. In the same way, a military combat robot does not really fight. It only emulates behavior that is alien to it.

Any such robot or AI operating in our physical human reality is condemned to be a stranger in a strange land, surrounded by things to which it cannot relate. We cannot expect a machine intelligence to truly understand uniquely human phenomena like warfare any more than we could understand the phenomena of a digital silicon world that evolved without human influence. Warfare may be one of the strangest and most unrelatable of all human behaviors. Warfare by its most basic definition—organized deadly violence against other groups within one's own species—is a custom that not even other earthly animals share with us, certainly not computers.

This is important because to the extent that we expect robots and AI to make more implicit or intuitive decisions in warfare, we will have to codify and embed into them the peculiar human rationale that guides us to fight and kill other members of our own species. As we bestow on military AI more autonomy and authority, we will require it to more broadly interpret this dangerous and ambiguous custom, which we ourselves do not always understand. The more we ask military AI to exhibit humanlike judgment, the more prone it will be to misinterpretation and to terrible errors.

THE BEAST WITHIN

Many of our fears about AI are based on projection. For instance, many fear that a hypothetical superintelligent, sentient AI may perceive humans as a threat and attempt to destroy us before we can harm it. That upon achieving "consciousness," it would be motivated by insecurity, existential panic, and the urge to power and choose violence in self-defense. Those fears essentially center on worries that the AI would act as we might be tempted to act in a similar situation. Such fears and expectations say more about our own species than they do about AI. On its own, AI is no more likely to want to kill than it is to experience gluttony for food or sexual desire. Those passions are utterly alien to an electronic computer. The only reason an AI might "think" and act that way is because humans trained it to do so. But that training could be accidental, such as through themes and patterns buried deep in the human-generated training data we provided it.

Future robots and AI may have lots of opportunities to infer bad lessons from human behavior. They will not have to look far. Plenty of examples show that the dark and savage impulse is still within us, beneath the surface. Scratch the surface, peel away the layers of theory and discipline, and the heart of darkness is often laid bare. The robots and AI that we create are dangerous because humans are. We are the beast in the machine.

In post-genocide Rwanda, peacekeepers and journalists found a rural village, in which nearly all the residents had been massacred. The body of a young boy of no more than ten lay in the midst, his neck half severed by the blow of a machete. His face was expressionless, but the wound grinned like a slice of bright melon, mocking human claims of reason and virtue. Looking around the sea of corpses a soldier might have muttered, "This isn't war." But it was, shorn of all pretense of civilization.

In a country that will remain nameless a small band of armed teenagers terrorized a town, committing murders and robberies. With no protection from the authorities, the frightened and enraged townspeople laid an ambush for the gang. Two males escaped but the townspeople caught a teenage girl who was with them. The mob beat and then burned her in the middle of a street. When the flames subsided she was charred but still moving, and a few of the crowd looked down uncertainly as if thinking, "What have we done? Should we help her?" Then someone dumped a pail of gasoline on her, and they had to step back from the fire as the crowd cheered its approval.

World War II in Europe did not start when German tanks crossed into Poland. It started years earlier when everyday German citizens accepted a genocidal vision that held that certain groups of humans are less than human, and that persecution and mass murder were the path to their own betterment. Before it appeared on the battlefield, the war was born in the nocturnal mob, the street riot, and the pogrom.

Savage tendencies aren't limited to foreigners. When emotions are hot they appeal to all humans. Fear, anger, and the desire to protect those we love, no matter what, tempt us to do things we might never consider in the calm light of day. The attitudes and behaviors we display from a position of safety can change quickly when that sense of safety is shattered.

Past crises show how quickly even American citizens and leaders can demand that the military throw aside concerns about proportionality and discrimination and "take the gloves off" in a fight with some adversary. Appeals such as "bomb them back to the stone age" become crowd-pleasers. Demands that the US military "take the gloves off" and use force without restraint appeared during the Vietnam War,[288] the post-9/11 war against al-Qaeda,[289] the war against the Taliban in Afghanistan,[290] and the war against ISIS.[291]

Despite their shortcomings, the professional militaries of modern democratic states in this century often serve as protective layers of lawful restraint, rationality, and discipline that moderate and modulate the urge to violence. They insulate us from the heart of darkness and protect us from ourselves. We may demand to bomb our enemies back to the stone age, but if we were to step into the cockpit of a bomber, we would face a confusion of dials and controls and have to ask our military to do it for us. The naval ship's combat information center and the nuclear missile silo are equally daunting and require the services of trained specialists. During their training those specialists will also have learned discipline, self-control, and the laws of armed conflict. They will have internalized the concept of military honor that requires, among other things, the sparing of noncombatants.

If robotics and AI make warfare too user-friendly, we may lose much of that insulation. Leaders responding to the chants of the crowd may decide of the military authorities: "We don't need you anymore, we can order it ourselves."

Modern war may have become highly technical, professionalized, and scientific. However, primal parts of our human brains such as the amygdala are still aroused, enraged, and animated by violence like nothing else. The primal brain can override reason and civilized habits. Heavyweight boxer Mike Tyson said that everyone has a plan until they get punched in the face. Violence is a most potent drug. Today, our reactions when we witness a terrorist attack, a mass shooting, or civil unrest may remind us that on some level the beast remains within us.

As social animals, humans have natural capacities for empathy and compassion. However, they are often reserved for those we regard as part of our own group. For those we perceive as outsiders or threats, we have unique capacities for callousness and brutality. Among the vertebrates only our closest cousins, the chimpanzees, engage in something like war, and only rarely, going on murderous rampages against clans whose territory conflicts with theirs. Researchers have witnessed mobs of male chimps surrounding members of other clans and beating them to death, and female chimps tearing apart infants with their hands.[292]

The violent animal spirits within us, like savage deities, are chained by civilization, but in a shallow dungeon not far below the surface. If we are driven by hatred, anger, or fear, or by the urge to protect loved ones or the desire for revenge, we might strike off the shackles of reason and legality that restrain them. And if in the future those primeval gods of war arise, they could rise into the robotic bodies that we have prepared for them.

In short, we should be careful what we wish for regarding military robots and AI that can act in a humanlike way. It will be profoundly difficult to produce a digital AI that can broadly interpret human intentions and emulate human behaviors in war. If we ever successfully create AI that truly acts as humans do, especially in violent crisis situations, we may regret the result.

Fortunately, for the near and midterm, the autonomy and authority of AI and robotic weapons are limited by immature technology. Current robotic systems cannot even maintain themselves in operation for long. But in the more distant future that could change. That is the path to altering not just the character of war—the weapons and tactics we use—but ultimately its nature as a human activity.

THE CHOICE WE MUST MAKE

It may be impossible to keep new powers from getting robotic weapons. But if humanity is to successfully navigate the dangers of these powerful rising military technologies and steer them toward an ethical path, predominance in robotic weapons must remain in the hands of ethical military authorities from nations that uphold human rights and the laws of war. If predominance passes to the many hungry disruptors, including reckless authoritarian states or violent non-state groups, then the risk of disasters will vastly increase.

Some concerned people may object and point out that democratic militaries, like those of the US, have committed abuses in the past. That is true. However, there are huge differences in how different governments and groups handle abuses and war crimes. There is no equivalence. For example, some US military members abused prisoners during the war in Iraq. As many as several dozen detainees may have died, either directly or indirectly, due to this abuse. Independent press reports and official investigations exposed the truth. I felt sickened when the news came out. It became a national disgrace, and rightfully so. Twelve prison guards were convicted of various charges. In contrast, Syrian officials tortured more than 14,000 detainees to death in Syrian dictator Bashar al-Assad's prisons during the Syrian civil war, and over 6,700 of their bodies are documented in photographs.[293] The Syrian government made no investigations. No perpetrators faced justice, at least not before Assad's overthrow at the end of 2024. Instead, those who exposed the atrocities had to flee for their lives. Those brutal deaths were not considered crimes or aberrations but the operation of state policy.

Similarly, some observers correctly point out that civilians have been killed in operations carried out by the militaries of democratic states. That is, sadly, still inevitable. However, there is no comparison with the wanton targeting of civilians conducted by hostile powers. Adversary forces' own actions demonstrate that they are aware of the difference. Many adversaries such as ISIS are so mindful of US and allied democracies' reluctance to harm noncombatants or civilian facilities that they regularly exploit that reluctance for their own military advantage through, for instance, the use of civilians as human shields.

If autocratic regimes such as in Russia, China, Iran, North Korea, or elsewhere that have no respect for human rights lead the world in robotic weapons and military AI, then our virtuous concerns about their ethics will become irrelevant. If non-state forces lead innovation or control the agenda, it may be worse. When ISIS invented the arming of small commercial hobby drones, it was also massacring tens of thousands of civilians and cutting off prisoners' heads on television.

Looking ahead, there are many thorny ethical questions remaining to be answered regarding the use of AI in combat. Where should the appropriate line be drawn between human control and judgment and AI authority for each military application? Are the moral boundaries different for systems that are inherently defensive, like active defense systems, and those with an attack capability? Should the line be different for systems that only destroy other unmanned systems than for those that may take human life, and is it possible to make that distinction?

The urgent moral question before us today is: Will the robotic military revolution be controlled mainly by those who agonize over ethical questions like these or by those who do not? Robotics and AI are unavoidably transforming warfare. We have frameworks that can address the complexities of applying them responsibly in military use. However, the path to the battlefield is much shorter for those who do not care for ethics. Unless we run quickly, they will get there first. Their rules will prevail, and they will impose them not through legal frameworks and international agreements but through the violent fait accompli and the rule of the gun.

10

THE WAY FORWARD

In addition to my service as a reserve officer supporting the US Air Force's science and technology enterprise, I have spent years in the private sector helping research-and-development-driven companies navigate strategic change. Guiding an industry leader through a turnaround to survive a wave of disruptive change is one of the toughest challenges in business. It has a lot in common with leading a military force that has just discovered it is on the verge of defeat to come back and win. In both cases, the toughest part can be overcoming the sense of paralysis caused by quickly changing circumstances.

Imagine the shock experienced by the CEO and board of Digital Equipment Corporation. In the late 1980s, the company was riding high as a leader of the mainframe computer industry, with 120,000 employees and record sales. It was one of the most profitable companies in the United States. Upstart personal computer makers were only a minor threat. Just two years later, it was losing money. The board fired the CEO, and soon half the company was laid off. Six years later, a surging PC maker, Compaq, bought what was left of the company. Once it began, change happened so quickly that the company's leaders couldn't figure out what to do.

In a similar way, the French generals at the start of World War II had been in a powerful position in Western Europe. Safe behind the massive high-tech defenses of the Maginot Line and possessing far more and better tanks than the German Army, they were confident. Yet shortly after the Germans attacked using their new blitzkrieg tactics, the French generals were shocked to learn that panzer divisions had already gotten behind them. They were paralyzed by situation reports that seemed to get worse by the hour. They could do nothing to stop their forces' collapse. Within weeks, France surrendered. The Armenian military in Nagorno-Karabakh suffered a similar experience. Once Azerbaijan's robotic offensive threw it into panic, it was almost impossible to change the course of events.

The best time to address disruption is before the crisis takes hold. Then, the dominant businesses or militaries often have more capabilities, talent, and money than the disruptors. In principle they should be able to respond effectively. Unfortunately, the category leaders are often complacent and fail to appreciate the developing threat. After the disruption becomes a crisis, they know they must act, but they cannot figure out what to do. The situation snowballs too quickly at that point for them to study it and assemble options.

While it has been awful, the war in Ukraine has helped to dispel the sense of complacency among today's leading democratic states. Their militaries can see the early effects of the oncoming revolution in robotic warfare on the battlefield and are seeking ways to respond. The challenge has shifted from acknowledging the threat to figuring out what to do about it.

Yet, barriers to action are inhibiting effective response. The financial resources of militaries such as the United States, Britain, and Japan seem large, but almost all their budgets are locked into covering bills they are already obligated to pay. Those include the operations and maintenance expenses for legacy equipment and bases, plus the costs of salaries and benefits for current and retired service members. Most of what is left for modernization is locked into major defense acquisition programs. They are committed to buying new versions of traditional platforms and to sustaining the massive industrial ecosystems that support them. The US Air Force and US Navy have cited budget constraints, in part, as reasons why they have fielded few unmanned aircraft that go beyond the Predator- and Reaper-style reconnaissance-strike drones of the War on Terrorism era.

The British Royal Air Force established a test unit to experiment with drones and uncrewed aircraft in 2020, but four years later it had not procured any drones or carried out any experiments. The UK minister for defense procurement admitted the unit had not been funded due to competing priorities.[294]

Demographic pressures are also inhibiting movement. The US military is trying to cover increasing global responsibilities with 40 percent fewer Navy and Air Force personnel than it had during Operation Desert Storm. Japan, with rapidly aging demographics, saw its prime recruiting-age population between eighteen and twenty-six years old drop by 38 percent between 1994 and 2021, and it has missed its military recruiting goals every year since 2014.[295] Britain has recently been forced to decommission Royal Navy warships ahead of schedule because it cannot recruit or retain enough sailors to crew them.[296] In the long term, embracing robotic warfare could relieve those demographic pressures. But in the meantime, today's urgent needs are pushing concerns about the future down the list of priorities.

Yet the evidence of recent conflicts shows that disruption is upon us. At that point, leaders want normative guidance that quickly answers the question: What should we do? Business leaders often call management consulting firms like the ones I worked for to provide those answers quickly. This chapter summarizes some normative guidance for quickly navigating the robotic revolution.

WHAT WE MUST DO

There are four basic imperatives for navigating the robotic revolution. We must immediately embrace first-wave robotic weaponry and the reality of the precision battlefield. Then we must quickly seize leadership in second-wave robotic systems. Third, we must lead in developing the strategies and tactics that capitalize on robotics and AI. And fourth, we must kick-start the transformation of the defense industrial base.

Immediately Embrace First-Wave Robotic Weaponry

As a recent US secretary of the Air Force put it, "We don't want to fight wars, we want to prevent them, and the way you prevent conflicts is to convince the other side that you have the will to resist and the capability to defeat aggression."[297] We must not create a window of vulnerability that tempts aggressors to use their new robotic weapons against us. We must foreclose the threat of adversaries conducting first-wave precision robotic warfare against friendly state militaries that are not prepared for it, as Azerbaijan surprised Armenia or as Iran and the Houthis surprised Saudi Arabia. We must field many new classes of inexpensive precision weapons and all that goes with them. That includes the complete kill web, including sensors, networking, and command and control. We must heed Maxim's warning, as discussed in chapter three, and accept that weapon-target asymmetry makes the large crewed platforms that form the core of our military power increasingly unlikely to survive in combat. Those platforms, and their crews and bases, dominate current military budgets, and we don't yet have the alternatives to replace them. So in the near term, we must make those platforms and bases bristle with as much active defense against small precision weapons as practical. Also, militaries must increase the lethal density of those platforms by adding large numbers of small precision weapons, particularly those that can engage targets beyond line of sight. Universal precision must become something our forces wield, not just something they face.

We must also embrace dispersion, camouflage, and concealment, including for what have traditionally been rear areas, such as bases. And we must expand capabilities such as electronic warfare. By accelerating these measures, leading democratic militaries can be ready for combat in the age of universal precision and delay the obsolescence of legacy platforms and forces.

Seize the Leadership in Second-Wave Robotic Systems

Making those first-wave adaptations will buy time. The goal must then be to reach and embrace the second wave as quickly as possible. That is the best opportunity for leading democratic states to reestablish lasting military advantage. The radical transformation of military platforms will enable militaries to restore maneuver and evade the oppressive grip of the precision battlefield. The Cambrian explosion of potential new robotic platforms offers limitless possibilities for "game changing" new concepts. Those that best embody the inherent tactical advantages of military robotics will rule the second wave.

Seizing the lead in the second wave requires being first to solve the challenges of dissociation. We must achieve key breakthroughs including control of large groups of elements, such as drone arrays, by a single operator; interoperability of elements; autonomous logistics; and seamless transition between networked and offline operation. For the US and other nations that require power projection, it may also involve the ability to integrate platforms for strategic mobility and then dissociate them for operations. We must make important advances in combat AI. Robotic platforms and dissociated units will deliver a wide range of game-changing new capabilities, such as runway-independent airpower. Those developments must emphasize the attributes that robotic combat history shows us are critical: low burden, low vulnerability, and efficient navigation in challenging environments. That implies extending autonomy through the complete mission cycle, including recovery and regeneration, employing intuitive manned-unmanned teaming, and favoring movement over the ground instead of on it.

Develop Strategies and Tactics That Capitalize on Robotics and AI

Military-technical revolutions happen when militaries combine emerging technologies with new methods of warfighting that fully capitalize on their inherent advantages. Being first to invent and apply those new warfighting methods is often more important than inventing the new equipment. We must develop new strategies and tactics for the robotic military revolution and then master their implementation in practice before our adversaries can. Control of the atmospheric littoral is one example. There will be more. Innovating radically new warfighting concepts requires creativity and free thinking, which plays to the strengths of democratic societies. But US and allied militaries must encourage and empower those efforts vigorously, as they did between the world wars. The US military has led many revolutions in warfighting methods, and we can do so again.

Kick-Start Transformation of the Industrial Base

Making those changes happen requires us to disrupt and rebuild our defense and national-security industrial base. Industry responds to changes in demand when the changes are expressed through actual buying behavior. Militaries must change their buying behavior to drive changes in industry. Much like the early aircraft industry, we can build the nascent stages of a bottom-up robotic weapons industry alongside existing military industry because it doesn't directly compete in similar product lines and doesn't cannibalize capacity. We must expand efforts to engage and cultivate new defense industry partners that have roots in leading commercial technology sectors such as mobile devices and software, including AI.

Because our current highly concentrated industrial base is locked into top-end products, new demand must focus on creating opportunities at the bottom of the market, targeting modest useful performance at a bargain cost. A target of at least 10x reduction in price below existing systems is necessary to drive the required paradigm shifts. A target of 100x reduction is more powerful and has already been achieved with the armed FPV drone. Expect the first generations of new systems to resemble the first PCs, useful but lower in performance than the top-end systems we are used to. But the very low prices will allow us to equip combat units that otherwise would have none—and in large quantities. The newly fostered industry base will improve them rapidly from that point. American industry has led many disruptive transitions in the global economy and can lead this one if the government helps to drive change.

BEWARE THE PATH OF LEAST RESISTANCE

In confronting these imperatives, we must remember that the leading democratic states are in the role of the disrupted, not the disruptors. It may take heroic effort for us to retain military predominance through the robotic revolution. We start with some important advantages, but they can slip away quickly. New competitors can evolve much faster. Disruption can be uncomfortable, but we must impose it on ourselves, or others will do it to us much more painfully.

Companies and militaries facing disruption often fail by defaulting to the path that is least disruptive to their existing operations and culture. US military operations and culture favor large, complex, and expensive systems. Networking more systems together is important to stave off competitors in the first wave, and we should capitalize on our starting advantages provided by investments in ISR, command and control, space systems, and other areas. However, we must not get stuck simply building more gigantic and exquisite networks of high-end systems.

Adding more nodes to a network increases its reach and power. Much like a social media network becomes more appealing and powerful the more people join it, there is power in connecting larger and larger networks of sensors, weapons, and platforms. But in the context of warfare, this approach has limits. An approach based on a contest to build larger and larger networks for delivering precision fires on an ever more massive scale may be unaffordable. It also exhibits top-down, mainframe-like thinking. It entrenches an attrition-warfare mindset and can prevent us from making the conceptual leaps that we require to achieve leadership in the second wave.

In addition, such networks will become too complex for human control and will have to be controlled with the help of centralized AI. The idea of an immense AI-enabled kill web is seductive but dangerous. The AI-controlled network approach can lead to network dependence and instabilities. New vulnerabilities and hazards may increase with scale even more quickly than new capability. A race to scale such military networks, and rely more and more on AI to manage them, encourages overempowerment and risks unforeseen consequences. As summarized by complex systems guru John Gall: "A complex system that works is invariably found to have evolved from a simple system that worked. A complex system designed from scratch never works and cannot be patched up to make it work. You have to start over, beginning with a working simple system."[298] Rigorous testing and employment of AI-controlled networks should start at a small scale and achieve much experience in the real world, before we contemplate scaling those networks beyond human control.

Lastly, this "least disruptive" approach embraces the kind of technological determinism that has repeatedly caused us to win battles but lose wars. Many American war-winning strategic concepts have promised victory. The only preconditions they require are superior technology, superior logistics, superior training, a superior industrial base, and a superior military budget. Even with all those preconditions in place, they have failed much of the time, such as in Vietnam, Afghanistan, and Iraq. A concept that can deliver victory without those preconditions would be truly powerful. Then our scale could multiply the advantage many times over. We need new concepts and strategies that apply the inherent advantages of robotics and AI to the challenges of actual twenty-first-century conflict, including the political realities of fourth-generation warfare. Those strategies might apply robotics and AI more effectively in a bottom-up manner, emphasizing better human-machine teaming at the tactical level.

SUPPORTING INITIATIVES

Enacting these imperatives will not be simple. But democratic militaries can leverage three supporting initiatives to accelerate the process. We should employ prototype units to cross the "valley of death" for disruptive concepts. We should promote change in military culture from within. And we should advance de facto norms for the use of military AI.

Crossing the "Valley of Death" with Prototype Units

Disruptive military innovations face a so-called "valley of death" that separates them from widespread adoption. This gulf of scrutiny, skepticism, and resistance is caused not only by inertia but also by the real unanswered questions and institutional gaps that can make it hard for a military to commit to a change that may be risky and expensive. The valley of death is as much cultural and organizational as technical. To cross it, the military often must develop new tactics and operational doctrine for applying innovations in combat that are just as radical as the new technology. It must train new kinds of warfighters who have the expertise in using the new technologies and methods. It may have to invent a new type of unit, with new functions. It might even have to cultivate new values and traditions that lead to creating a new warfighting community within the service.

In Nagorno-Karabakh in October 2020, an Armenian tank attempting to hide under a tree is targeted by the laser designator of an Azerbaijani drone. (Azerbaijan Ministry of Defense)

Traditional defensive lines were of little use in Nagorno-Karabakh. Here, an Armenian armored vehicle behind a protective earthwork is targeted by another Azerbaijani drone. (Azerbaijan Ministry of Defense)

Robotic weapons are increasingly prominent on the global market. The Emirati defense company EDGE Group displays its unmanned systems at the UMEX defense exhibition in Abu Dhabi in 2022. (EDGE Group / Wikipedia)

Technicians power up a Kettering Bug, one of the first aerial torpedoes or cruise missiles, in preparation for a test flight circa 1918. (National Museum of the US Air Force)

Two Ferdinand tank destroyers and their supporting Borgward B IV remote-controlled breaching vehicles prepare to advance before the battle of Kursk in July 1943. (Rowman & Littlefield Publishing Group / Heinz Henning)

US Navy personnel service a TDR-1 assault drone during the deployment of STAG-1 to the Pacific Theater in 1944. The first-person-view television camera in the nose is exposed for servicing. (US Navy / Naval History and Heritage Command)

German infantry soldiers prepare Goliath remote-controlled demolition vehicles for an assault during the Warsaw Uprising in August 1944. (Bundesarchiv, Bild 101I-695-0411-06 / Josef Gutermann)

The light cruiser USS *Savannah* takes a direct hit from a German Fritz X radio-guided precision anti-ship bomb off Salerno, Italy, on September 11, 1943. (US Naval Heritage and History Command / NH 95562)

A B-61 Matador pilotless nuclear bomber (later known as the TM-61) is ready for launch during a field exercise in the 1950s. (National Museum of the US Air Force)

The Firebee drone “Tom Cat” survived 68 missions over Vietnam. The versatility and effectiveness of Firebee drones in combat zones prompted advocacy for remotely piloted combat aircraft in the 1970s. (National Museum of the US Air Force)

Masses of spent artillery shell casings near British artillery batteries in World War I illustrated the woeful inefficiency of unguided industrial-age weapons. (National Museum of Scotland / Tom Aitken)

Bomb craters from US attempts to destroy a V-1 launch site in Beauvoir, France, in 1944. Ninth Air Force bombers attacked the site 12 separate times. The target is a structure within the wooded area in the center right of the image. (National Museum of the US Air Force)

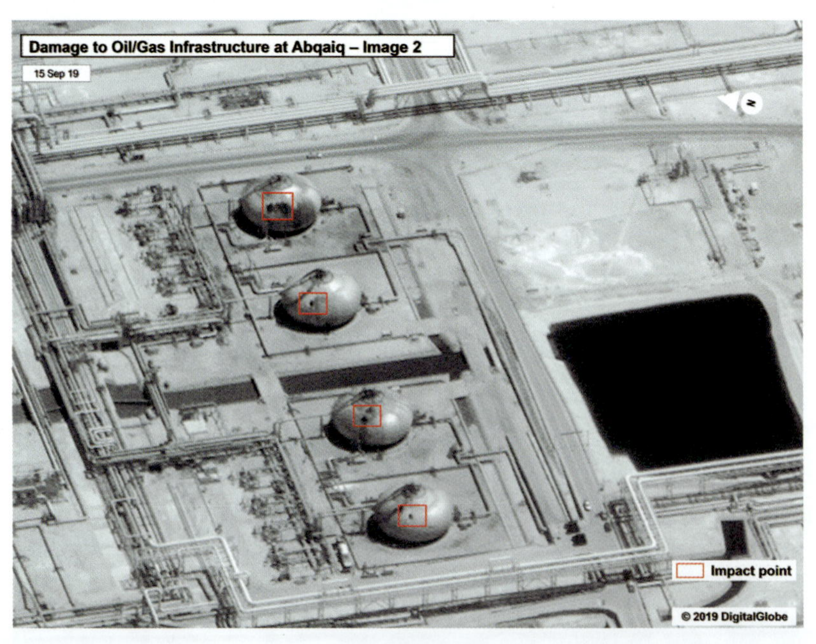

Precision guidance increases the lethality of weapons by 100–1,000 times. This satellite photo shows four of the hits on the Saudi Arabian oil facility at Abqaiq inflicted during the counter-value strike by Iranian / Houthi kamikaze drones on September 14, 2019. (US Government / DigitalGlobe via AP)

In dense urban battlegrounds, strikes by large precision-guided weapons produce great amounts of collateral damage, as shown by the aftermath of an airstrike in Gaza in November 2023. (AFP via Getty Images)

First-person view (FPV) drones combine a small video drone with an unguided munition such as this rocket-propelled grenade warhead to produce a small, cheap, and effective precision weapon. (Wojciech Grzedzinski for *The Washington Post* via Getty Images)

In Ukraine, precision weapons destroyed thousands of Russian armored vehicles. Here, destroyed Russian vehicles litter the terrain following a failed armored attack in Vuhledar, in Donetsk Oblast, south of Avdiivka. (Armed Forces of Ukraine/Tavriya Operational-Tactical Group)

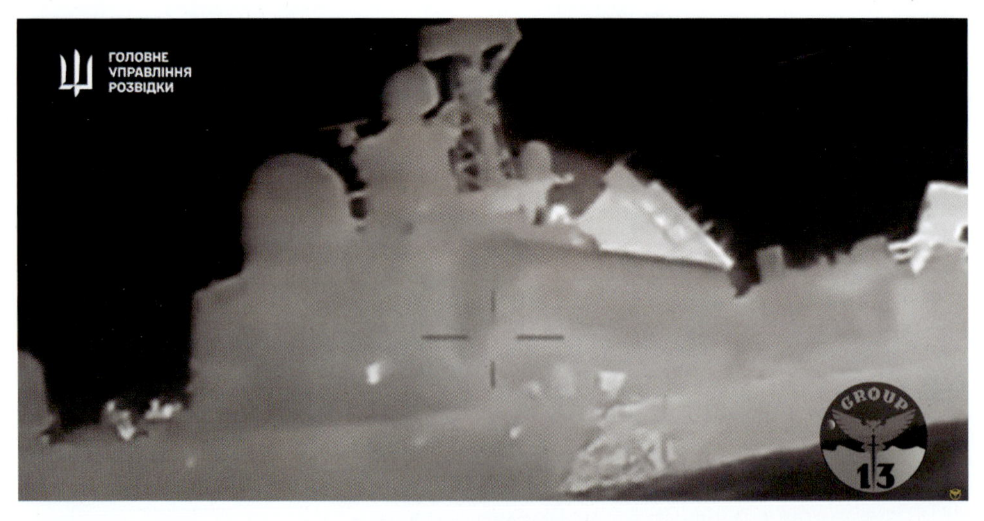

A first-person-view night-vision image from a Ukrainian MAGURA v5 drone boat attacking the Russian warship *Ivanovets* on February 1, 2024. The hole left by the earlier strike of another drone boat is visible at the waterline. (Defense Intelligence of Ukraine)

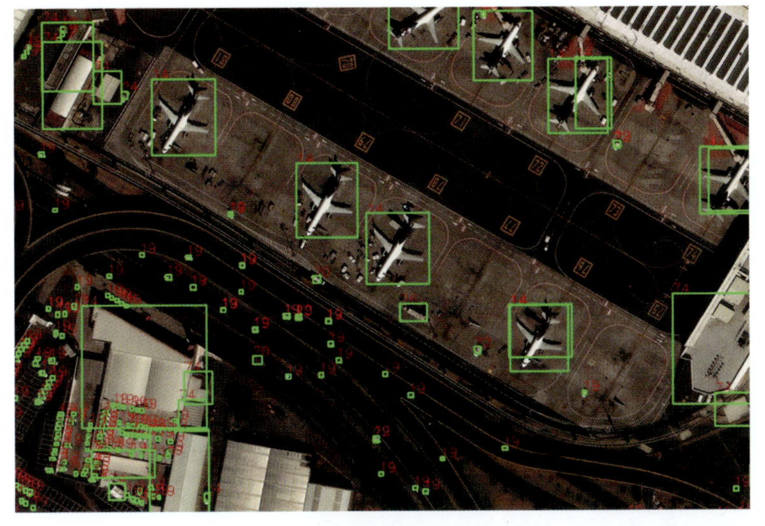

Deep learning AI can quickly find and identify targets such as aircraft in video imagery, pointing the way toward highly selective signature-seeking weapons. (Defense Innovation Unit / Nikolay Sergievskiy and Alexander Ponamarev)

An XQ-58 Valkyrie unmanned aircraft flies in formation with an F-16 during a test flight in 2023, an early example of the "loyal wingman" concept of manned-unmanned teaming in the air domain. (US Air Force / Master Sergeant Tristan McIntire)

The Black Hornet micro-unmanned air vehicle (UAV) weighs 18 grams, illustrating the state of the art in miniaturization for robotic military platforms. (US Air Force / Senior Airman Miranda Mahoney)

The Tactical High Power Operational Responder (THOR) is a prototype high-power microwave system for defense against future drones and drone swarms. (US Air Force Research Laboratory / John Cochran)

The RACER Heavy Platform (RHP), a prototype of potential future unmanned ground vehicles, undergoes autonomous navigation and mobility testing in April 2024. (Defense Advanced Research Projects Agency)

Four-legged mobile robots or “robot dogs” are increasingly used to assist with security tasks such as patrolling base perimeters, complementing human security police and military working dogs. (US Department of Defense / Senior Airman William Pugh)

Military experiments are developing techniques for the control of large numbers of drones, foreshadowing the rise of future swarming tactics. (US Department of Defense / Aaron Duerk)

A vertical takeoff and landing drone prepares to launch from the flight deck of an Arleigh Burke-class destroyer. Runway independence will be increasingly important for robotic aircraft in combat. (US Department of Defense / Petty Officer 2nd Class Elliott Schaudt)

Two Saildrone unmanned surface vessels operate with an Arleigh Burke-class destroyer in the Persian Gulf in 2023. Removing the need for human crew will enable many novel designs for future military platforms. (US Department of Defense / Navy Petty Officer 2nd Class Jeremy Boan)

Heavy drones currently under test, such as this Griff 135 with a 66-pound (30-kg) payload capacity and 45-minute endurance, approach the performance needed by combat drones in the atmospheric littoral. (US Air Force Research Laboratory / Samuel King Jr.)

Illuminated drones deploy in array formation during a show in Busan, South Korea, on January 1, 2024. While pre-scripted, these formations foreshadow future maneuver swarms for combat operations in the atmospheric littoral. (@kimjaelyong via YouTube)

Human psychological biases play a major role in human-machine interaction. Popular images of robots and AI used in advertising and publicity often depict them with a pleasing humanlike face, using the bias toward anthropomorphism to establish positive expectations in the human viewer. (Pixabay / patrypguerrero)

The spectrum of military operations includes many missions other than high-intensity combat, such as humanitarian assistance. The most useful robotic systems must be flexible and relevant to many different missions. (US Department of Defense / Staff Sergeant Jason Bushong)

Successful modern warfighting involves much more than destroying targets. The human and political dimensions of war require expert human judgment, and the decisions of junior leaders can have strategic consequences. (US Department of Defense / Staff Sergeant Sierra Melendez)

Manufacturing modern integrated military platforms requires great investment in industrial infrastructure and expertise. Here, F-35 fighters are assembled in the Lockheed-Martin facility in Fort Worth, Texas. (Lockheed-Martin / Chris Hanoch)

Many militaries use deliberate attacks against civilian targets as a warfighting strategy. Russian strikes using guided missiles destroyed this shopping mall and residential complex in Kyiv in March 2022. (Aris Messinis for AFP via Getty Images)

Military working dogs have many abilities, such as the ability to track the gaze of a human handler, that enable them to interpret human intent. Similar capabilities will be important for future military robots designed for human-robot teaming. (US Department of Defense / Taylor Curry)

The well-established teaming relationship between human warfighters and military working dogs offers a successful model that can help guide ethical and effective human-robot teaming in combat. (US Air Force / Staff Sergeant Daniel Asselta)

Clayton Christensen observed that the practices that make good sense in assessing and adopting sustaining innovations can inhibit the adoption of disruptive innovations. He called this the "innovator's dilemma." Militaries can embrace sustaining innovations like improved tanks or missiles or fighter planes because they have experience conducting operations with similar systems. The units that will be customers for a particular innovation know what qualities they want the improved technology to have. They understand how it fits into their established doctrine. In addition, they are ready to use their influence within the military service to advocate for it.

In contrast, for a disruptive innovation, the technology may be unfamiliar and immature, the requirements may not be well defined, and the operational doctrine may not exist. That combination of unknowns can make it very hard to satisfy the normal criteria to commit to an innovation. Complicating matters, there is no existing constituency within the service to champion it. Fortunately, the US military has a proven, but nearly forgotten, solution to that problem.

In the business world, startup companies often take disruptive technologies across the "valley of death." Small and nimble, they can quickly experiment and iterate to answer the open questions and find the right technology, business model, and early customers to generate success in the market. Large companies have difficulty entering small and risky new markets because their operations are too large and expensive to allow them to experiment and quickly change direction as they figure out how to succeed. However, Christensen pointed out that the most innovative large companies create small, startup-like units to make and sell the innovation on an experimental basis.[299] Those "internal startups" can identify the customers, rapidly iterate to find out what the market wants, and discover and validate a profitable business model on a small scale. Then, once they have answered those key questions, the resources of the company can flow in to scale the solution.

The US military has a strong history of doing the same thing by creating prototype units. It has used prototype units to blaze the trail for nearly every important disruptive military innovation of the twentieth century, from carrier aviation and mechanized forces to stealth attack and long-range reconnaissance-strike drones.

Disruptive Innovation	Service	Prototype Unit (Year Established)
Carrier Aviation	US Navy	USS *Langley* (1922)
Mechanized Forces	US Army	1st Cavalry Regiment (Mechanized) (1933)
Helicopter Aviation	USMC	HMX-1 (1947)
ICBMs	USAF	576th Strategic Missile Squadron (1958)
Air Assault	US Army	11th Air Assault Division (Test) (1965)
Stealth Attack	USAF	4450th Tactical Group (1979)
Unmanned Reconnaissance-Strike	USAF	32nd Expeditionary Air Intelligence Sq. (2000)

Examples of the prototype units that enabled the US military services to successfully field prior disruptive military innovations.

A prototype unit advances and clarifies the technology, requirements, and operational doctrine through the process of building a first-of-its-kind operational unit employing a disruptive innovation. Like a startup, it uses rapid iteration to discover and solve the many problems of implementing a disruptive new capability in practice. It results in a deployable and combat-ready military unit that serves as a model for the future units of its type. It also builds the culture and constituency to protect and grow the new method of warfighting.

For example, the Predator started out as a long-endurance ISR platform, and early test deployments to Bosnia and Kosovo in the late 1990s showed the immense potential of real-time aerial surveillance.[300] On August 1, 2000, the Air Force established a new provisional unit, the 32nd Expeditionary Air Intelligence Squadron, to use modified Predators to hunt for high-value al-Qaeda targets in Afghanistan in partnership with the CIA.[301] They modified the Predators with global satellite datalinks that allowed operators to control them from anywhere on Earth. Flying from Uzbekistan, but controlled from ground stations in Germany and later from CIA headquarters in Langley, Virginia, the unit found targets that very likely included Osama bin Laden.[302] However, without weapons, there was no good way for the military to act on the real-time information. Therefore, the unit returned to the US and improvised a way to arm its Predators with Hellfire missiles. After the 9/11 attacks, the Air Force reactivated the unit as the Air Combat Command Expeditionary Air Intelligence Squadron on September 17, 2001.[303] Using those Hellfire missiles, it soon conducted the first armed drone strikes against individual Taliban and al-Qaeda targets. Because it had figured out the appropriate equipment, organization, and techniques to conduct reconnaissance-strike operations, the squadron became the model for the many similar Air Force units that followed.

The US military created most prototype units during peacetime. They worked out the details through intense field exercises, which often pitted them against conventional units. They used rapid iteration, incorporating the lessons from exercises and experiments and adjusting accordingly. The requirement to field an actual fighting unit forced teams of technical and operational experts to solve the many problems along the way to pull the innovation across the "valley of death."

Many prototype units were deployed into combat. For instance, in 1965, after two years as a prototype unit for helicopter air assault, the Army renamed the 11th Air Assault Division (Test) as the 1st Cavalry Division (Air Mobile). It became the first complete Army division sent to Vietnam and almost immediately entered combat operations including the Battle of Ia Drang in November 1965.[304] Similarly, the Air Force's 4450th Tactical Group, the secret prototype unit that refined the design and operations for the F-117 stealth attack jet during the 1980s, was rebadged as the 37th Tactical Fighter Wing and had dramatic combat impact during Operation Desert Storm.[305]

Prototype units serve as great partners for smaller innovative companies. They can adapt and iterate at similar speed. By filling orders for small numbers of systems to equip prototype units as they evolve, the companies receive the revenue and realistic customer feedback they need to quickly mature and grow.

The Cambrian explosion of robotic military concepts will demand that the military and industry explore and mature many new concepts. The US Air Force is leveraging its history with prototype units to field a new experimental operations unit for unmanned Collaborative Combat Aircraft—or loyal wingmen.[306] The other military services can do the same. Using the powerful method of building prototype units will allow the military to quickly cross the "valley of death" in emerging areas such as atmospheric littoral operations and others that are yet to be conceived.

Change Military Culture from Within

Militaries naturally try to honor and preserve the practices and culture that won past victories. Yet if they maintain old traditions and ways of fighting too rigidly, they can become resistant to change, especially disruptive change.

For example, the United States developed the first low-cost precision loitering munition in the 1980s, but it was never deployed because of change-resistant culture within the US Army.[307] The FOG-M missile system provided a miniature vertical launch array packed with precision non-line-of-sight missiles. Each missile carried a video camera in its nose that transmitted a live first-person view to the launch vehicle over a thin optical fiber that spooled out behind the missile for several kilometers. It was immune to jamming. A soldier could use the camera to search and then fly the missile into any target, including tanks, fortifications, and flying helicopters. It was cheap and gave a simple infantry vehicle like a Humvee the power to destroy an entire company of enemy armor from a safe distance. Despite successful demonstrations, it did not fit neatly into established categories of weapons. The Army could not decide whether it should belong to the infantry, artillery, or air defense branches. So it canceled the program. The US Army lost decades of opportunity as a result. It bought essentially the same system almost forty years later in the form of the Israeli Spike NLOS missile, which the Azerbaijani Army used in Nagorno-Karabakh. When future robotic weapons are developed, they are likely to confound traditional categorization. We must be ready to upset conventional thinking or be left behind.

Military leaders should encourage this process through words and actions. They can also encourage junior military personnel to question traditional habits in areas that are subject to technological disruption. There are a myriad of overlooked opportunities to remove complacent habits.

As just one example, current military training exercises are supposed to stress our systems and warfighters, but the way they are planned can unintentionally reinforce the obsolete assumption that we will choose the time and place to fight. Navies, for instance, should move away from testing shipboard active defenses against missile or drone attacks during scheduled exercises when the crews are alert and ready. Those produce artificially high performance and a false sense of security. The deadly attacks increasingly come unexpectedly. For more realism, simulated attacks could come by surprise when a ship is sailing to or from the exercise area—or even when it is docked in port.

Establish de Facto Ethical Norms for Military AI

As the guarantors of the international rule of law, the US and other democratic states must continue to establish and uphold legal or "de jure" constraints such as the laws of war, the law of the sea, and the Geneva Protocols. However, those laws rest on the authority of states and their multilateral organizations. Powers that seek to disrupt the international rules-based order do not care about following international rules. We should not expect de jure rules to carry much weight unless we can enforce them with real, de facto power.

When the Syrian government broke the prohibition against using chemical weapons against civilians in 2018, democratic states enforced the prohibition not by threat of charges in an international court but by destroying Syria's chemical weapons facilities with long-range precision munitions. In a similar way, Houthi precision-weapon strikes on commercial shipping in violation of the laws of the sea were addressed by a twenty-nation naval coalition that intercepted the Houthis' missiles and drones and destroyed military sites in Yemen. The Cold War prohibition against a nuclear strike remained unbroken for decades not because a nuclear strike would be illegal but because it would result in immediate nuclear retaliation.

If we are to establish and enforce norms regarding military robotics and AI, democratic states must master them first to secure our military advantage. Only then can we set responsible examples and precedents and have the leverage to enforce those norms until they are globally observed. Justice only has consequence while there is power to enforce it. Otherwise, as Thucydides observed almost 2,500 years ago, "the strong do what they can and the weak suffer what they must."

In the past, most proposed ethical norms for robotic weapons focused on the specific case of counterterrorism drone strikes during the War on

Terrorism, or narrowly on the issue of human control, or else on simplistic proposals to ban so-called "killer robots." More recently, experts are advancing proposals that are more broadly applicable and technically informed.[308]

Frameworks like the ones advanced in earlier chapters offer a more comprehensive and rigorous approach to "how" to achieve strong ethical use. To keep a clear focus on what really matters, we should also anchor the desired outcomes, or the "what," on real-world effects, which are observable and measurable. Ethical judgments should focus on measurable, outcome-based performance metrics like selectivity, not on how they are achieved. That keeps the emphasis on the ethical capability of the whole system, including the contribution of the non-robotic parts of the system. More advanced technology that produces worse outcomes is not progress. Focusing our standards on outcome-based metrics will also help us keep our ethical norms relevant as robotic and AI technologies rapidly evolve.

For example, consider how we reduced traffic accidents at intersections. At first, human traffic police directed the vehicles at each intersection using flags and whistles. That became impractical as the numbers of cars and trucks multiplied. Therefore, police departments installed the first electric traffic signals. At first, police officers operated those signals from a vantage point using manual switches. Over time the signals became more automated, and the police officers' oversight became more distant. Today all traffic lights are controlled by computers without any direct human oversight.

The important goal in that case was not that all traffic lights had human oversight, it was a low rate of traffic accidents. That was measurable. It made sense for a human officer to supervise the early primitive electrical signals because they were too unreliable to deliver the desired outcome for long without supervision. But as the technology improved, they became able to deliver it reliably on their own. Today, nobody would suggest that police officers supervise traffic lights.

Combat AI is of course far more complicated than traffic control. However, traffic lights are safety critical, and lives are at stake if they fail. In other safety-critical areas such as medicine, we also use performance metrics like selectivity and efficacy as the means to assess safety, and we base our regulations on them. We should use the same approach to anchor our de facto norms about military robotics and AI.

These three initiatives—employing prototype units, promoting change in military culture from within, and developing de facto norms for the use of military AI—can help the militaries of democratic states more quickly establish leadership in the robotic revolution. They will also help anchor the coming changes in the ethical norms and values professed by the militaries of the free world.

Despite all the factors we have discussed, it may seem at times as if we are drifting toward a future where the role of robotics and AI in warfare assumes the model of a centralized AI-controlled robotic kill web. In that model, a military networks its sensors and weapons together, hands over control of anything from a tactical engagement to the entire war to an opaque and brittle machine intelligence, pushes the button, and hopes that we can live with the results. There will certainly be situations where rapid, AI-coordinated action is required. However, there will be many others where that isn't appropriate. That default "AI controls all" model is a trope from decades of science fiction. Despite being a cliché, it can sometimes seem as if it is the only model on offer for thinking about the future of human warfighters, robots, and AI. But that isn't true.

A HOPEFUL MODEL

We have a successful example that may guide our way toward a future of more effective and responsible military human-machine teaming. Tyndall Air Force Base sits amid a humid coastal pine forest on the Florida panhandle. Severely damaged by a hurricane in 2018, the Air Force is rebuilding it into its showcase "installation of the future."

The base is protected by the 325th Security Forces Squadron. Its personnel are leading experts at employing an intelligent autonomous system that has proven its effectiveness and resilience in combat in multiple conflicts across the spectrum of military operations. That remarkable system is the military working dog. Speaking with the squadron's leaders including Major Jordan Criss and Master Sergeant Andre Hernandez, I was struck by their reverence for these special animals that accompany human warfighters into danger. The squadron is also the Air Force's lead unit for integrating robot dogs.[309] On a typical day, an airman on the base might notice a security police officer calling commands to a military working dog while a robot dog trundles by, whirring and clicking on its patrol route. The unit is the Air Force's leading authority on the relative abilities of both systems.

A long history with military working dogs reveals a powerful model for effective human-machine teaming in the real environment of war. The United States Air Force has been the lead service for training US military working dogs since 1952.[310] Over time, military working dogs progressed from being used for limited tasks under very close control to being more and more autonomous, mirroring the more recent progress of military robots. In the early years, the Air Force used them essentially as sensor platforms, helping sentries to guard air bases and other installations. They helped extend the radius of presence of their handlers. Early studies showed that a single sentry teamed with a trained dog could watch as much perimeter as ten sentries standing guard.[311]

During the Vietnam War, military dog handlers experimented with new roles, such as accompanying infantry on patrol. Then, during the War on Terrorism, the science of employing military working dogs advanced tremendously. They became integral parts of some combat teams. Their tasks became much more varied and challenging, requiring the dogs to operate farther from their handlers, in unpredictable environments. Many worked "on point," screening the advance of human soldiers on patrol, searching for hidden enemies, explosives, or booby traps. Others accompanied special forces on raids and missions to apprehend high-value human targets. They have become multimission assets, extending the presence and effect of human warfighters in a wide range of situations. Their military handlers have learned how to develop and employ higher degrees of autonomy in these dogs while maintaining appropriate control.

The methods that handlers use to direct military working dogs resemble those that engineers are seeking to develop for the control of future military robots. They use short commands, sometimes code words, that communicate directions like "stay next to me," "search," "bite and hold," "release the object," and "return to me." They are simple and intuitive, and the handler can give them while doing other tasks. Their simplicity takes advantage of the dog's intelligence to interpret them in each situation and decide the details of how to comply.

In addition, dogs take an active role in communication. They interpret their handler's expressions, body language, and tone of voice, often detecting nonverbal signals of which the handler may not be conscious. A military working dog will follow its handler's gaze and infer where his or her attention is focused. It learns each handler's unique mannerisms. As Cameron Ford, one of the world's leading experts in military working dog training, puts it, "the dog learns the handler faster than the handler learns the dog."[312]

A well-trained working dog applies contextual reasoning and intuition to properly interpret its handler's intent. For instance, if one suspect in a group starts to run, and the handler commands "bite and hold," the dog understands intuitively that this refers to the person who is running away. If an expert dog receives such a command before entering a room on a raid, but it finds the room full of children or American soldiers, it will use its own judgment to abort without biting anything.[313]

This allows a dog to accept a simple form of mission command. While handlers strive to maintain what they call "positive control," many missions require a dog to operate autonomously, such as searching a structure or apprehending a distant suspect. A highly trained dog understands what mission it is on and can stay focused on it for long periods. For instance, Navy SEAL handlers have released dogs on search-and-apprehension missions to catch individuals who escaped high into rugged mountains. The dogs often find and catch the suspect up to half an hour later and miles away.[314] Instead of its autonomy introducing uncertainty, the dog uses its autonomy to overcome unforeseen obstacles and complications and more certainly achieve the outcome desired by its handler.

Military working dogs exhibit mental flexibility that contrasts with the brittleness of today's AI. Trainers select new working dogs for their high intelligence. Through native abilities that are enhanced by training, dogs can disregard unfamiliar elements and distractions in the most challenging and unfamiliar environments. A chaotic situation like a raid may be a riot of unfamiliar settings and objects, loud noises, and strange odors. Dogs can't be oblivious to those surroundings because they may have to consider them in deciding their actions, but they can intuitively distinguish what is important from what is not and focus on their task. In one test, a trainer in a padded suit was challenged to hide inside a building from a dog that would try to find and apprehend him. The trainer found a shower room far from the entrance, turned on the showers to create a curtain of distraction to hide behind, and turned off all the lights so that it was pitch black. Pleased with the cleverness of his hiding place, he was surprised when shortly thereafter the dog came blasting straight through the darkness and falling water, ignoring all the unfamiliar elements, knocked him down on the slick floor, and held him.[315]

Military working dogs exhibit other abilities that we seek from combat AI. They know when they are under attack. They eat, drink, and sleep autonomously. They can pass an object permanence test in which a dog must remember where a handler has hidden a toy despite misdirection and find it again immediately. That basic skill is important in many tactical situations. Dogs also incorporate new experiences with their previous training and generalize to apply their training in new contexts, something beyond the capacity of current deep-learning AI. When circumstances prevent them from doing something the way they were trained, they improvise an alternative.

In one test, an experienced dog was sent to apprehend, knock down, and hold a trainer in a bite suit who was waiting with another trainer high on a mountainside. The trainer bet his colleague that the dog could not knock him down. A walkie-talkie call announced that the dog had been released in the valley far below. The trainers waited fifteen minutes. Then the dog appeared, racing up the slope toward them like a guided missile. The dog went straight for its target, ignoring the second trainer. The expert trainer hunched over and braced himself. Observing the trainer, at the last second the dog swerved from the direct path and circled a clump of bushes to appear behind the trainer. Catching him in mid-turn and off balance, the dog knocked the trainer flat and held him while his colleague laughed.[316]

Dogs have emotional intelligence. Expert dogs know what they are doing and don't like to be used unethically. A dog's demeanor and body language give feedback to its handler that a command he or she gave might pose a problem. It may hesitate or look at its handler in discomfort. Where dogs were once used in questionable roles such as interrogation and crowd control, regulations now prohibit those uses. If future AI-powered robots obtain a similar understanding of context, they could help notice questionable situations and assist their military handlers to avoid accidents or misuse.

The 325th Security Forces Squadron's experience with robot dogs confirms their usefulness while highlighting their differences from living dogs.

Major Criss and Master Sergeant Hernandez explain that while robot dogs have legs for mobility over rough terrain, they are not dogs at all. A robot dog is a sensor platform that works like a ground-based drone to provide coverage of a wide area, like the original use of sentry dogs decades ago. They compare the robots to "base defense Roombas."[317] The robots get a patrol path they can walk autonomously while sending an on-the-ground video feed to the base operations center. They send an alert if they see something anomalous, but they don't need to be watched every minute, which allows the human security forces to monitor a large area.

The military working dogs have much greater agility and can do many more things than the robots. No one expects today's robot dogs to swim across a lake or race over a vertical wall on an obstacle course the way military working dogs can. However, the robots provide unique capabilities. They do not become tired, hungry, cold, or agitated. They never become bored or upset. They transmit real-time video, and a security forces defender can speak through the robot to a human intruder to scare them or to occupy them while human defenders are on the way. Letting robots do what they do well saves both dogs and humans for what they do best: handling the more demanding situations that need judgment and emotional intelligence, such as de-escalating situations when a human is caught entering the base.

Most importantly, robot dogs are attritable. The security forces squadron can send them to do things they would never ask of a dog or a human defender. For instance, when surprising an invading adversary force, the adversary will focus intensely on the first display of gunfire. While the shooter draws the adversary's attention, other friendly responders can maneuver relative to it. No human responder wants to be the one to draw that attention from the enemy. However, if an armed robot could absorb that initial reaction, it gives a lot of options to the other responders.[318]

Robot dogs possess limited AI. While nowhere near as self-sufficient as real dogs, the robots can plug themselves in to recharge, and they are slowly reducing their burden and vulnerability. They are also getting better at navigating difficult environments like tall grass. While the security forces can control them from a great distance, today their remote control is simple and explicit. As Cameron Ford points out, there is nothing yet in human-robot teaming that compares to the powerful, intuitive two-way bond between a military working dog and its handler. However, we could strive to emulate that in future human-robot interfaces.[319] For instance, research is starting to enable robots to interpret human gestures and facial expressions.

Military working dogs provide a compelling model for effective human-machine teaming. However, while they possess many abilities far beyond those of current robots and AI, no one expects military working dogs to understand our wars or to fight our battles for us. As living things, they are more than just weapons. However, we calibrate our expectations of them. Thus, they also provide a sanity check on tendencies toward magical thinking and other biases toward overempowerment.

We can direct many years of fruitful development toward giving military robots and AI some of the abilities that make military dogs so effective in the messy real world. We will have to adapt and translate some of those abilities into the very different missions and domains in which future military robots will operate. Robots that work in the air and at sea may require tailored methods. However, human experiences handling other animals such as dolphins and falcons may provide similar insights. We will also need to extend the methods from one-on-one interactions to the control of dissociated swarms such as drone arrays. Whatever the application, experience with military working dogs can teach us a lot about how we can build a trusted working relationship with a nonhuman intelligence, and how we could build robots that are designed to team more efficiently with us.

It's no coincidence that since the days of the early visionaries of robotics, the developers and operators of military robots have tended to name their robots after animals, most often dogs. John Hays Hammond and Benjamin Miessner started that trend with their electric dog in 1912. The German remote-controlled panzer units and American assault drone units in World War II both used the dog metaphor, and it has been repeated many times since. That tendency might have been telling us something all along. The relationship between human warfighters and their working dogs exemplifies the trust that we can dream of one day having with robots. Teamed together, handlers and their dogs can do things that neither can do alone. Each provides unique abilities, and their intuitive connection prepares them to handle the unexpected.

Seeing a handler and a working dog returning to base against a Florida sunset, one can glimpse a positive future. The beast that we embed into the robots that we take to war need not be a dark one from our own nature. It could be another trustworthy and humane one, for which a successful model already exists. By building on the principles that have provided success with military working dogs, we may find a path to developing and employing military robotics and AI in ways that are both effective in the real world of combat and that reflect the humanity and sanity we hope to bring to future conflicts.

THE FUTURE IS UP TO US

Management guru Peter Drucker said, "The best way to predict the future is to create it."[320] We can't deny the advance of robotics and AI and their inevitable applications to human conflict. However, we can anticipate the challenges that are coming and guide developments in positive directions. Making that happen is up to all of us. The impacts of the robotic military revolution go far beyond the military and the battlefield and will affect us all. So, we all have a stake. And we all have a voice, at least in democratic countries.

Politicians, the media, and citizens can all take action. Politicians should help give the military the flexibility to transform faster. That includes the ability to shift resources and to divest expensive aging assets like obsolete bases. As the makers of laws and regulations, they can promote ethics that are based on principles and frameworks that are rigorous enough to deliver the outcomes we desire. They can also take foreign policy actions that demonstrate our commitment to our principles, helping to establish de facto norms that position robotics and AI as tools that implement, not circumvent, the laws of war.

The media should move beyond scary stories about the rise of "killer robots." We are past the stage of raising questions and making admonishments and in the stage of finding answers and implementing practical solutions. News organizations and cultural media have huge roles to play in moving the popular debate beyond simple hand-wringing about the entire subject and focusing it on finding the more substantive solutions that this subject deserves.

Citizens can insist, using our votes, clicks, and views, that politicians and the media address these urgent issues. The first and most important thing that citizens can do is be well informed. If the people are well informed and understand the urgency for action, politicians, the media, and democratic militaries will follow.

We have the grounding to guide the future wisely. We understand the major trends that will determine supremacy on future robotic battlefields. We can see how robotic warfare fits within the strategic challenges of achieving political success in wars as they are fought in the twenty-first century. We also have the frameworks we need to recognize and avoid the most dangerous risks of AI as applied to combat. By applying the tools of successful disruptive military innovation to the challenge of the robotic revolution, we can outmaneuver our adversaries both old and new. We can build a future where robotics and AI serve to secure a global community that is based on respect for human rights and the rule of law.

The robotic military revolution is happening now. There is no time to lose if we want to determine the future course of its evolution and its impacts on our world and avoid the worst effects of disruption. But there is time to act, if we move quickly.

Robotics and AI were born with military applications in mind. The founding inventors' vision over a century ago was to use them to oppose aggression and reduce senseless suffering and destruction, thereby making war less uncertain and more rational, discriminate, and ethical. Since then, progress has largely validated that vision. The robotic transformation of combat is only beginning. If we act, we can help determine whether its positive potential instead of its dangers will be realized. We can weather the tempest of disruption, and perhaps, if we are up to the task, deliver to future generations an even less violent and savage world, which may, as Tesla once hoped, serve as a preliminary to permanent peace.

ACKNOWLEDGMENTS

This book is the product of many years of work, and it is a pleasure for me to acknowledge the contributions of many people along the way. First, I want to acknowledge the patient and often unsung help of many librarians, archivists, and curators. Their kind and selfless assistance was essential, particularly in helping me to uncover rare records and piece together the often poorly documented history of robotic warfare and the powerful lessons that it offers. I am particularly indebted to Brett Stolle at the archives of the National Museum of the US Air Force. His support was instrumental in helping to germinate the seed of my research in its earliest stages beginning in 2006. I thank Don Sando, Ted Maciuba, Rick Heaton, and others at the US Army Maneuver Center of Excellence who were early champions of my concept of the atmospheric littoral and helped to showcase my ideas within the Army starting in early 2021. Much appreciation also to John Antal, Angelina Callahan, and many other military forward thinkers, for stimulating conversations and encouragement. Thanks to Dan Ward, Mario Serna, Brian Fry, Adam Goobic, and Jordan Criss, as well as my wife, Vanesa Sanchez, each of whom reviewed draft chapters and provided their expert military, scientific, or literary insights. Special thanks to my agent, Jud Laghi, as well as my editor Rick Chillot, publisher Glenn Yeffeth, and the rest of the amazing team at BenBella Books, whose invaluable support and feedback helped to bring the book to a high polish. Most of all, thanks to Vanesa, whose unwavering patience, love, and support made this book possible.

NOTES

INTRODUCTION

1. Can Kasapoglu, "ANALYSIS—Five Key Military Takeaways From Azerbaijani-Armenian War," *Anadolou Post*, October 30, 2020, https://www.aa.com.tr/en/analysis/analysis-five-key-military-takeaways-from-azerbaijani-armenian-war/2024430.
2. "The Fight for Nagorno-Karabakh: Documenting Losses on the Sides of Armenia and Azerbaijan," *Oryx*, September 27, 2020, https://www.oryxspioenkop.com/2020/09/the-fight-for-nagorno-karabakh.html.
3. Andriy Dubchak, "Лабораторія саморобних бомб для FPV," Frontliner (YouTube Channel), April 4, 2024, https://m.youtube.com/watch?v=1S-ok4is-GM.
4. Andrew S. Grove, *Only the Paranoid Survive: How to Exploit the Crisis Points That Challenge Every Company and Career* (New York: Doubleday, 1996).
5. "The Air Force We Need: 386 Operational Squadrons," Secretary of the Air Force Public Affairs press release, September 17, 2018, https://www.af.mil/News/Article-Display/Article/1635070/the-air-force-we-need-386-operational-squadrons/.
6. "Drone Warfare," The Bureau of Investigative Journalism, April 2020, https://www.thebureauinvestigates.com/projects/drone-war.
7. Colin Demarest, "One-Third of U.S. Military Could be Robotic, Milley Predicts," *Axios*, June 11, 2024, https://www.axios.com/2024/07/11/military-robots-technology.

8. Robert Work, Reagan Defense Forum Speech on the Third Offset Strategy, Reagan Presidential Library, Simi Valley, CA, November 7, 2015.
9. Umar Farooq, "The Second Drone Age: How Turkey Defied the U.S. and Became a Killer Drone Power," *The Intercept*, May 14, 2019, https://theintercept.com/2019/05/14/turkey-second-drone-age/.
10. Farooq, "The Second Drone Age."
11. Joseph Trevithick, "Chinese Flying Wing UCAV Testing Accelerating Based on Satellite Imagery, Videos," *The War Zone*, September 5, 2024, https://www.twz.com/air/chinese-flying-wing-ucav-testing-accelerating-based-on-satellite-imagery-videos.
12. Joseph Trevithick, "China's New Stealthy Trimaran Drone Ship: Our Best Look Yet," *The War Zone*, November 9, 2024, https://www.twz.com/news-features/our-best-look-yet-chinas-new-stealthy-trimaran-drone-ship.
13. Pablo Robles, "China Plans to Be a World Leader in Artificial Intelligence by 2030," *South China Morning Post*, October 1, 2018, https://multimedia.scmp.com/news/china/article/2166148/china-2025-artificial-intelligence/.
14. Robles, "China Plans to Be a World Leader in Artificial Intelligence by 2030."
15. Sun Chi, "China's Investment in AI Expected to Reach $38.1b in 2027," *China Daily*, August 23, 2023, https://global.chinadaily.com.cn/a/202308/23/WS64e5b34fa31035260b81dc9f.html.
16. Joseph Trevethick, "North Korea Unveils Clones of Israeli Kamikaze Drones," *The War Zone*, August 26, 2024, https://www.twz.com/air/north-korea-unveils-clones-of-israeli-kamikaze-drones.
17. "Putin: Leader in Artificial Intelligence Will Rule World," Associated Press, September 1, 2017, https://apnews.com/bb5628f2a7424a10b3e38b07f4eb90d4.
18. Sagi Cohen, "Gaza Becomes Israel's Testing Ground for Military Robots," *Haaretz*, March 3, 2024, https://www.haaretz.com/israel-news/2024-03-03/ty-article-magazine/.premium/gaza-becomes-israels-testing-ground-for-remote-control-military-robots/0000018e-03ed-def2-a98e-cfff1e640000.
19. Sudarsan Raghavan, "In Libya, Cheap, Powerful Drones Kill Civilians and Increasingly Fuel the War," *Washington Post*, December 22, 2019, https://www.washingtonpost.com/world/middle_east/libyas-conflict-increasingly-fought-by-cheap-powerful-drones/2019/12/21/a344b02c-14ea-11ea-bf81-ebe89f477d1e_story.html.

20. "Pakistan Seen Deploying Unmanned Ground Vehicle Near Indian Border," *DefenseWorld.net*, July 6, 2020, https://www.defenseworld.net/news/27355/Pakistan_Seen_Deploying_Unmanned_Ground_Vehicle_Near_Indian_Border__Media_Reports#.XxIbS5uSnIU.
21. Sebastien Roblin, "Israel's Kamikaze Drones Are Causing Problems for Syria," *National Interest*, November 11, 2019, https://nationalinterest.org/blog/buzz/israels-kamikaze-drones-are-causing-problems-syria-95051.
22. David Axe, "Turkey Has a Drone Air Force. And It Just Went to War in Syria," *National Interest*, March 2, 2020, https://nationalinterest.org/blog/buzz/turkey-has-drone-air-force-and-it-just-went-war-syria-128752.
23. "Timeline: Houthis' Drone and Missile Attacks on Saudi Targets," *Al Jazeera*, September 14, 2019, https://www.aljazeera.com/news/2019/09/timeline-houthis-drone-missile-attacks-saudi-targets-190914102845479.html.
24. Christopher Cavas, "New Houthi Weapon Emerges: A Drone Boat," *Defense News*, February 19, 2017, https://www.defensenews.com/digital-show-dailies/idex/2017/02/19/new-houthi-weapon-emerges-a-drone-boat/.
25. Robert Postings, "The Islamic State's Armed Drone Program," *International Review*, December 10, 2017, https://international-review.org/islamic-states-armed-drone-program/; Ben Watson, "The Drones of ISIS," *Defense One*, January 12, 2017, https://www.defenseone.com/technology/2017/01/drones-isis/134542/.
26. "Slaughterbots," Autonomousweapons.org, 2017, https://autonomousweapons.org/.
27. Raymond Kurzweil, *The Singularity Is Near* (New York: Viking, 2005).
28. *Opere Inedite di Francesco Guicciardini III: Storia Fiorentina* (Firenze: Barbera, Bianchi and Company, 1859), 105.

CHAPTER 1

29. James J. Hall, *American Kamikaze* (Titusville, FL: J. Bryant, Ltd., 1984), xiii; Laurence R. Newcome, *Unmanned Aviation: a Brief History of Unmanned Aerial Vehicles* (Reston, Virginia: American Institute of Aeronautics and Astronautics, 2004), 69.
30. Nikola Tesla, "My Inventions," *Electrical Experimenter*, February–October 1919 (Serialized), http://www.tfcbooks.com/e-books/my_inventions.pdf.

31. Nikola Tesla, "The Problem of Increasing Human Energy," *Century Magazine*, June 1900, 175–211.
32. Nikola Tesla, "Method of and Application for Controlling Mechanism of Moving Vessels or Vehicles," US Patent 613809, filed 1 July 1898, and issued 8 November 1898, https://pdfpiw.uspto.gov/.piw?Docid=613809.
33. Nikola Tesla, "The Problem of Increasing Human Energy."
34. Benjamin F. Miessner, *Radiodynamics: The Wireless Control of Torpedoes and Other Mechanisms* (New York: D. Van Nostrand Company, 1916), 199.
35. Lawrence Sperry, "The Aerial Torpedo," *U.S. Air Services* 11, no.1, January 1926, 16–19.
36. George O. Squier, "Automatic Carrier for the Signal Corps (Liberty Eagle)," letter to the chief of staff, October 5, 1918.
37. Reginald Bacon, *The Dover Patrol, 1915–1917, Vol. I* (New York: George Doran, 1919), 211–214, https://books.google.com/books?id=ceBCAAAAIAAJ.
38. Angelina Callahan, "Reinventing the Drone, Reinventing the Navy: 1919–1939," *Naval War College Review* 67, no. 3, 2014, 98–122.
39. Samuel J. Cox, "Forgotten Valor: USS *Pioneer* (AM-105) and the sinking of HMT *Rohna*, the Worst Loss of U.S. Life at Sea, 26 November 1943," H-Gram 022, Appendix 2, Naval History and Heritage Command, October 2018, https://www.history.navy.mil/content/history/nhhc/about-us/leadership/director/directors-corner/h-grams/h-gram-022/h-022-2.html.
40. David Irving, *The Rise and Fall of the Luftwaffe: The Life of Field Marshal Erhard Milch* (London: Focal Point Publications, 1973), 259.
41. Frank W. Heilenday, *V-1 Cruise Missile Attacks Against England: Lessons Learned and Lingering Myths from World War II*. Report P-7914 (Santa Monica, California: RAND Corporation, 1995), 6; Michael J. Armitage, *Unmanned Aircraft* (London: Pergamon/Brassey's Defence Publishers, 1988), 17.
42. Weike [Commander, Panzer-Abteilung (Fkl) 300], "Summary of Battalion Experiences at Sevastopol," June 22, 1942, reproduced in Markus Jaugitz, *Funklenkpanzer: A History of German Army Remote- and Radio-Controlled Armor Units*, trans. David Johnston (Winnipeg, Manitoba: J. J. Fedorowicz Publishing, 2001), 109–110; Inspector General of the Armed Forces, "Interim Guidelines for the Employment of Radio-Controlled Armored Vehicles," Berlin, April 2, 1943, reproduced in Jaugitz, *Funklenkpanzer*, 594–598.
43. Christopher W. Wilbeck, *Sledgehammers: Strengths and Flaws of Tiger Tank Battalions in World War II* (Bedford, PA: Aberjona Press, 2004), 71; Wolfgang

Schneider, *Tigers in Combat I* (Winnipeg, Manitoba: J. J. Fedorowicz Publishing, 2000), 147.

44. Delmar S. Fahrney, "The Genesis of the Cruise Missile," *Aeronautics & Astronautics*, January 1982, 34–39, 53; Newcome, *Unmanned Aviation*, 67.
45. Oscar Smith, "Special Air Task Force—Training Task Force and Project Option—Accomplishments and Final Report," memorandum to US Fleet and chief of naval operations, Enclosure C, December 7, 1944. (US National Archives); Nick T. Spark, "Unmanned Precision Weapons Aren't New," *Proceedings of the U.S. Naval Institute* 131, no. 2, February 2005, 66–71.
46. T.W. South II, "War Diary—1 September 1944 to 1 November 1944," memorandum, Special Task Air Group One, United States Fleet, October 31, 1944, reproduced in James J. Hall, *American Kamikaze* (Titusville, FL: J. Bryant, Ltd., 1984), 201–206.
47. Claude Larkin, "Operations of STAG ONE Detachment in Northern Solomons Area, Report On," memorandum, Commander Aircraft, Northern Solomons, October 30, 1944, reproduced in Hall, *American Kamikaze*, 205–208.
48. Trent Hone, "Countering the Kamikaze," *Naval History* 34, no. 5, October 2020, https://www.usni.org/magazines/naval-history-magazine/2020/october/countering-kamikaze.
49. John S. McCain, Sr., "Assault Aircraft Drone (TDR-1 and TD3R) Program: Present Status of and Recommendation Concerning," memorandum, US Navy, April 19, 1944 (US National Archives); Fahrney, "The Genesis of the Cruise Missile."
50. Antony L. Kay, *Buzz Bomb* (Boylston, Massachusetts: Monogram Aviation Publications, 1977), 8.
51. Theodore von Kármán, "Where We Stand: A Report Prepared for the AAF Advisory Group," Dayton, Ohio: Headquarters Air Materiel Command, January 1946, 14.
52. Mark C. Cleary, *6555th Test Wing: Missile and Space Launches through 1970*. 45th Space Wing History Office, US Air Force, November 1991, 25.
53. Russell Hawkes, "Subsonic Snark Adds Effectiveness to SAC Forces," *Aviation Week*, September 15, 1958, 50–64.
54. Duncan Lennox, "SM-62 Snark," in *Jane's Strategic Weapon Systems* (electronic database), Jane's Information Group, UK, September 30, 2004.
55. Darrel Whitcomb, "PAVE NAIL: There at the Beginning of the Precision Weapons Revolution," *Air Power History* 58, no. 1, Spring 2011, 14–27.

56. US Air Force, *History USAF Drone/RPV*, Wright-Patterson Air Force Base, Ohio: RPV System Program Office, 1976, 10–11; William Wagner, *Lightning Bugs and Other Reconnaissance Drones* (Fallbrook, CA: Aero Publishers, 1982), 182–183.
57. Wagner, *Lightning Bugs and Other Reconnaissance Drones*, 157–165, 176.
58. "Remotely Flown Vehicles Stir Wide Interest," *Aviation Week & Space Technology*, June 14, 1971, 26; Wagner, *Lightning Bugs and Other Reconnaissance Drones*, 186–189.
59. Terrell E. Greene, "The Rise of Remotely Manned Systems," *Astronautics & Aeronautics*, April 1972, 44–53.
60. William B. Graham, "RMVs in Aerial Warfare," *Astronautics & Aeronautics*, May 1972, 36–47.
61. Benjamin S. Lambeth, "AirLand Reversal," *Air Force*, February 1, 2014, https://www.airforcemag.com/article/0214reversal/.
62. David A. Deptula, *Effects-Based Operations: Change in the Nature of Warfare* (Arlington, Virginia: Aerospace Education Foundation, 2001), 2.
63. Walter J. Boyne, "How the Predator Grew Teeth," *Air Force* 92, no. 7, July 2009, 42–45.
64. Alec Bierbauer and Mark Cooter, *Never Mind, We'll Do It Ourselves: The Inside Story of How a Team of Renegades Broke Rules, Shattered Barriers, and Launched a Drone Warfare Revolution* (New York: Skyhorse Publishing, 2021), 274–295.
65. John M. Koetz and Bruce E. Brendle, "Project Remote: Desert Storm Robotics," *Personal Perspectives of the Gulf War* (Arlington, VA: Institute of Land Warfare, Association of the United States Army, 1993), 101–103.
66. James Schmitt, "No More Stovepipes: Unifying Air Operations Planning Doctrine to Enable Multirole Mission Success," *Wild Blue Yonder*, July 23, 2023, https://www.airuniversity.af.edu/Wild-Blue-Yonder/Article-Display/Article/3447567/no-more-stovepipes-unifying-air-operations-planning-doctrine-to-enable-multirol/; John D. Duray, *Remotely Piloted Aircraft Operations: Lessons Learned and Implications for Future Warfare*, Mitchell Forum report No. 28 (Arlington, VA: Mitchell Institute for Aerospace Studies, December 2019), 14, https://mitchellaerospacepower.org/remotely-piloted-aircraft-operation-lessons-learned-and-implications-for-future-warfare/.

67. Tyler Rogoway, "The Alarming Case of the USAF's Mysteriously Missing Unmanned Combat Air Vehicles," *The War Zone*, July 2, 2020, https://www.thedrive.com/the-war-zone/3889/the-alarming-case-of-the-usafs-mysteriously-missing-unmanned-combat-air-vehicles.
68. *National Defense Authorization Act for Fiscal Year 2001*, Public Law 106-398, sec. 220, October 30, 2000.

CHAPTER 2

69. Michael Gilbert, *The Somme: Heroism and Horror in the First World War* (New York: Owl Books, 2006), 37.
70. Augustin M. Prentiss, *Chemicals in War: A Treatise on Chemical Warfare* (New York and London: McGraw-Hill, 1937), 660, 662.
71. L. Van Loan Naisawald, "The Cost in Ammunition of Inflicting a Casualty," Technical Memorandum ORO-T-246, Chevy Chase, Maryland: Johns Hopkins University Operations Research Office, July 28, 1953.
72. Boyd L. Dastrup, *King of Battle: A Branch History of the US Army's Field Artillery* (Ft. Monroe, Virginia: US Army Training and Doctrine Command Historian, 1992), 213.
73. Marilyn M. Harper, *World War II and the American Home Front* (Washington, DC: National Park Service, US Department of the Interior, 2007), 3, https://irma.nps.gov/DataStore/downloadfile/465955.
74. General William. C. Westmoreland, speech at Tufts University, Medford, Massachusetts, December 12, 1973.
75. Norman Friedman, *Naval Firepower: Battleship Guns and Gunnery in the Dreadnought Era* (Barnsley, UK: Seaforth, 2008), 166.
76. Jeffrey R. Barnett, *Future War: An Assessment of Aerospace Campaigns in 2010* (Maxwell AFB, Alabama: Air University Press, 1996), 11; David A. Deptula, *Effects-Based Operations: Change in the Nature of Warfare* (Arlington, Virginia: Aerospace Education Foundation, February 2001), https://www.airforcemag.com/PDF/DocumentFile/Documents/2005/EBO_deptula_020101.pdf.
77. *The United States Strategic Bombing Surveys* (Maxwell AFB, Alabama: Air University, October 1987), 13, https://apps.dtic.mil/dtic/tr/fulltext/u2/a421958.pdf.

78. Barnett, *Future War*; Deptula, *Effects-Based Operations*.
79. Richard P. Hallion, "Bombs That Were Smart Before Their Time," *World War II Magazine*, September 2007, 52–57.
80. Edward Westerman, *Flak: German Anti-Aircraft Defenses, 1914–1945* (Lawrence, Kansas: University Press of Kansas, 2001), 293–294.
81. Peter J. Burke, "XM1156 Precision Guidance Kit (PGK)," 52nd Annual Fuze Conference, Sparks, Nevada: National Defense Industrial Association, 13–15 May 2008, https://ndiastorage.blob.core.usgovcloudapi.net/ndia/2008/fuze/VABurke.pdf; Geneva International Center for Humanitarian Demining (GICHD), "Explosive Weapon Effects—Final Report" (Geneva, Switzerland: GICHD, February 2017), 34, https://www.gichd.org/fileadmin/GICHD-resources/rec-documents/Explosive_weapon_effects_web.pdf.
82. Michael Peck, "The US Has Given Ukraine Nearly 1 Million 155 mm Artillery Shells. Now It's Looking for US Companies to Build More of Them." *Business Insider*, September 13, 2022, https://www.businessinsider.com/us-wants-to-build-artillery-shells-as-it-supplies-ukraine-2022-9.
83. Jeff Schogol, "Russia is Hammering Ukraine with up to 60,000 Artillery Shells and Rockets Every Day," *Task and Purpose*, June 13, 2022, https://taskandpurpose.com/news/russia-artillery-rocket-strikes-east-ukraine/; Jon Jackson, "Russia Has Lost Half Its Combat Capability in Ukraine: U.K. Defense Chief," *Newsweek*, July 7, 2023, https://www.newsweek.com/russia-has-lost-half-its-combat-capability-ukraine-uk-defense-chief-1811042.
84. Timothy J. Sakulich, *Precision Engagement at the Strategic Level of War: Guiding Promise or Wishful Thinking?* Occasional Paper no. 25 (Maxwell AFB, AL: Air War College, December 2001), 11; Jack Sine, "Defining the 'Precision Weapon' in Effects-Based Terms," *Air and Space Power Journal* 20 (1) 2006, 81–88.
85. Kyle Mizokawi, "The CIA's Blade-Wielding 'Flying Ginsu' Missile Strikes Again," *Popular Mechanics*, July 28, 2019, https://www.popularmechanics.com/military/weapons/a30175425/cia-blade-missile/.
86. Malcolm Gladwell, *The Bomber Mafia: A Dream, a Temptation, and the Longest Night of the Second World War* (New York: Little, Brown and Company, 2021), 45–46.
87. Phillip S. Meilinger, *10 Propositions Regarding Airpower* (School of Advanced Airpower Studies, US Air Force History and Museums Program, 1995), https://media.defense.gov/2010/May/25/2001330281/-1/-1/0/AFD-100525-026.pdf.

88. US Army Acquisition Support Center, "Excalibur Precision 155 mm Projectiles," October 11, 2018, https://asc.army.mil/web/portfolio-item/ammo-excalibur-xm982-m982-and-m982a1-precision-guided-extended-range-projectile/; Tim Mahon, "Technology: The Final Frontier: EXCALIBUR—Coming at You from Every Direction," *Military Technology*, April 2019, 53.
89. Udi Etzion, "IDF Shooters Get 'Smart' Gun Sight to Increase Accuracy," *Ynetnews.com*, January 13, 2019, https://www.ynetnews.com/articles/0,7340,L-5444185,00.html.
90. Tyler Rogoway, "This Portable Remote Weapon Turret Is Right Out of Call of Duty," *The Drive*, July 22, 2022, https://www.thedrive.com/the-war-zone/35031/this-portable-remote-weapon-turret-is-like-contra-and-call-of-duty-video-games-come-to-life.
91. T. N. Dupuy, "Quantification of Factors Related to Weapon Lethality (Annex III)," *Historical Trends Related to Weapon Lethality* (Washington, DC: Historical Evaluation and Research Organization, October 15, 1964), H5, https://apps.dtic.mil/sti/pdfs/AD0458759.pdf.
92. Dupuy, "Quantification of Factors Related to Weapon Lethality," H13–H16.
93. Dupuy, "Quantification of Factors Related to Weapon Lethality," H29–H33.
94. Saadia Amiel, "Defensive Technologies for Small States," in Williams, Louis, ed., *Military Aspects of the Israeli-Arab Conflict* (Tel Aviv: University Publishing Projects, 1975), 13–58.
95. Westerman, *Flak: German Anti-Aircraft Defenses*, 294.
96. Alfred W. Johnson, *The Naval Bombing Experiments Off the Virginia Capes—June and July 1921* (Washington, DC: Naval Historical Foundation, 1959), 14, https://www.history.navy.mil/research/library/online-reading-room/title-list-alphabetically/n/the-naval-bombing-experiments.html.
97. US Government Accountability Office, *F-35 Joint Strike Fighter: Development Is Nearly Complete, but Deficiencies Found in Testing Need to Be Resolved*, GAO-18-321 (Washington, DC: GAO, June 2018), https://www.gao.gov/assets/gao-18-321.pdf.
98. Jeffrey A. Vish, "Guided Standoff Weapons: A Threat to Expeditionary Air Power," thesis, Naval Postgraduate School, 2006, 39, https://apps.dtic.mil/dtic/tr/fulltext/u2/a457379.pdf.
99. Howard Altman, "Moment of Drone Strike That Destroyed Russian Il-76s Seen in Infrared Image," *The War Zone*, August 31, 2023, https://www.twz.com/moment-of-drone-attack-that-destroyed-il-76s-at-russian-base-seen-in-infrared-image.

100. Sara Fritz and Karen Tumulty, "'Smart Bombs' on Target at Air HQ Post: Arms: Lasers Guided Explosives with Unprecedented Accuracy, Boosting Role of U.S. Technology," *Los Angeles Times*, January 19, 1991, A9, https://www.latimes.com/archives/la-xpm-1991-01-19-mn-201-story.html.
101. MBDA Missile Systems, "Sea Venom / ANL," MBDA Systems, 2020, https://www.mbda-systems.com/product/sea-venom-anl/.
102. Mizokawi, "The CIA's Blade-Wielding 'Flying Ginsu' Missile Strikes Again."
103. Chad Stahelski, dir. *John Wick: Chapter 2*. Los Angeles, CA: Summit Entertainment, 2017, https://www.youtube.com/watch?v=rsuNowyCF0c.
104. Andrew F. Krepinovich and Steven M. Kosiak, "Smarter Bombs, Fewer Nukes," *Bulletin of the Atomic Scientists*, November/December 1998, 26–32.
105. Deptula, *Effects-Based Operations*, 2001.
106. US Joint Chiefs of Staff, *Joint Publication 3-0: Joint Operations* (Washington, DC: US Department of Defense, October 22, 2018), III–30.
107. Merel A.C. Ekelhof, "Lifting the Fog of Targeting: 'Autonomous Weapons' and Human Control Through the Lens of Military Targeting," *Naval War College Review* 71, no. 3, Summer 2018: Article 6, https://digital-commons.usnwc.edu/cgi/viewcontent.cgi?article=5125&context=nwc-review.
108. US Office of the Director of National Intelligence, *The AIM Initiative: A Strategy for Augmenting Intelligence Using Machines*, Washington, DC: January 16, 2019.
109. Robert Wall, "The Devastating Impact of Sensor Fuzed Weapons," *Air Force Magazine*, March 1, 1998, https://www.airforcemag.com/article/0398sensor/.
110. Jacob Shermeyer, Thomas Hossler, Adam Van Etten, Daniel Hogan, Ryan Lewis, and Daell Kim, "RarePlanes: Synthetic Data Takes Flight," 2021 IEEE Winter Conference on Applications of Computer Vision (WACV), January 5–9, 2021, 207–271.
111. Simon K. Mencher and Long G. Wang, "Promiscuous Drugs Compared to Selective Drugs (Promiscuity Can be a Virtue)," BMG Clinical Pharmacology 5, no. 3, 2005, https://bmcclinpharma.biomedcentral.com/track/pdf/10.1186/1472-6904-5-3.pdf.
112. John Spencer, "Why Militaries Must Destroy Cities to Save Them," Modern War Institute, November 8, 2018, https://mwi.usma.edu/militaries-must-destroy-cities-save/.

CHAPTER 3

113. Azerbaijani Ministry of Defense, "Cəbhənin Xocavənd istiqamətində düşmənin 'Tor-M2KM' ZRK-sı vurulub," November 9, 2020, 1:43, https://www.youtube.com/watch?v=C0pcbeSm0Sw.
114. "Turkish Air Force Used E-7A Peace Eagle to Hunt and Destroy S-300 Using TB2 Drones," Global Defense Corp, November 7, 2020, https://www.globaldefensecorp.com/2020/11/07/turkish-air-force-used-e-7a-peace-eagle-to-find-and-destroy-s-300-using-tb2-drones/.
115. Andrew E. Kramer and Anton Troianovski, "Azerbaijan and Armenia Agree to Cease-Fire in Nagorno-Karabakh," *New York Times*, late edition (East Coast), October 9, 2020.
116. John Antal, *7 Seconds to Die: A Military Analysis of the Second Nagorno-Karabakh War and the Future of Warfighting* (Philadelphia: Casemate Publishing, 2022), 3.
117. John Antal, *7 Seconds to Die*, 79.
118. Merrill A. McPeak and Robert A. Pape, "Hit or Miss," *Foreign Affairs* 83 no. 5, September/October 2004, 160–163.
119. Hilaire Belloc and Basil Temple Blackwood, *The Modern Traveller* (London: Edward Arnold, 1898), 41.
120. William S. Murray, "Revisiting Taiwan's Defense Strategy," *Naval War College Review* 61, no. 3, Summer 2008, Article 3; Charles Bronk and Gabriel Collins, "Bear, Meet Porcupine: Unconventional Deterrence for Ukraine," *Defense One*, December 24, 2021, https://www.defenseone.com/ideas/2021/12/bear-meet-porcupine-unconventional-deterrence-ukraine/360195/.
121. Francis Farrell, "Record Russian Armor, Personnel Losses in Failed Attempt to Take Avdiivka by Storm," *Kyiv Independent*, October 25, 2023, https://kyivindependent.com/none-of-it-made-any-sense-understanding-russias-disastrous-offensive-on-avdiivka/.
122. "18 Feb: Shocking Footage Reveals the Real Cost Russians Paid for Avdiivka," Reporting from Ukraine, February 18, 2024, https://www.youtube.com/watch?v=VNTC0H1zWwI.
123. "Update from Ukraine, 01/28/2024," Denys Davydov, January 28, 2024, https://www.youtube.com/watch?v=37_mBXpjXeQ.
124. Ellie Cook, "Russia's Staggering Avdiivka Losses Laid Bare by Ukraine," *Newsweek*, February 18, 2024, https://www.newsweek.com/russia-losses-equipment-tanks-avdiivka-ukraine-1870987.

125. Luke Harding, "Cheap but Lethally Accurate: How Drones Froze Ukraine's Frontlines," *The Guardian*, January 25, 2024, https://www.theguardian.com/world/2024/jan/25/how-drones-froze-ukraine-frontlines.

126. James J. Schneider, "The Theory of the Empty Battlefield," *RUSI Journal* 132, no. 3, 1987, 37–44.

127. John C. Schulte, *An Analysis of the Historical Effectiveness of Anti-Ship Cruise Missiles in Littoral Warfare* (Monterey, California: Naval Postgraduate School, 1994), 35.

128. Steve Brown, "ANALYSIS: The Magura-V5 Sea Drone—Scourge of Russia's Black Sea Operations," *Kyiv Post*, March 6, 2024, https://www.kyivpost.com/analysis/29068.

129. "Ukrainian Saboteurs Behind Attacks Inside Russia, Reports Say," Voice of America News, August 22, 2023, https://www.voanews.com/a/russia-downs-ukrainian-drones-in-moscow-region-/7234981.html; Peter Suciu, "Ukrainian Video Shows Destruction of Il-76 Aircraft—Contradicting the Kremlin's Claims," *Forbes*, September 1, 2023, https://www.forbes.com/sites/petersuciu/2023/09/01/ukrainian-video-shows-destruction-of-il-76-aircraft--contradicting-the-kremlins-claims/?sh=3e4f29318f4b.

130. T. S. Rowden, *Surface Force Strategy: Return to Sea Control.* (San Diego, CA: U.S. Naval Surface Forces Pacific Fleet, 2016); Edward Lundquist, "DMO Is Navy's Operational Approach to Winning the High-End Fight at Sea," *Seapower*, February 2, 2021.

131. US Air Force. *Air Force Doctrine Note 1-21: Agile Combat Employment* (Maxwell Air Force Base, AL: Curtis E. LeMay Center for Doctrine Development and Education, 2022), https://www.doctrine.af.mil/Portals/61/documents/AFDN_1-21/AFDN%201-21%20ACE.pdf.

132. David Oliver, "Ukraine's Unmanned Air War," *Armada International*, June/July 2022, 26–28.

133. Jack Watling and Nick Reynolds, *Meatgrinder: Russian Tactics in the Second Year of Its Invasion of Ukraine* (London: Royal United Services Institute, 2023), 18.

134. David Hambling, "Ukraine Wins First Drone vs. Drone Dogfight Against Russia, Opening a New Era of Warfare," *Forbes*, October 14, 2022, https://www.forbes.com/sites/davidhambling/2022/10/14/ukraine-wins-first-drone-vs-drone-dogfight-against-russia-opening-a-new-era-of-warfare.

135. Howard Altman, "Ukraine Bracketing Key Kursk Highway with Drones to Slow Russian Logistics," *The War Zone*, August 28, 2024, https://www.twz

.com/news-features/ukraine-bracketing-key-kursk-highway-with-drones-to-slow-russian-logistics.

136. Sam Schechner and Daniel Michaels, "Ukraine Has Digitized Its Fighting Forces on a Shoestring," *Wall Street Journal*, January 3, 2023, https://www.wsj.com/articles/ukraine-has-digitized-its-fighting-forces-on-a-shoestring-11672741405.
137. Watling and Reynolds, *Meatgrinder: Russian Tactics in the Second Year of Its Invasion of Ukraine*, 13–14, 24.
138. Joe Barnes, "How Dummy HIMARS are Depleting Russia's Missile Supplies," *The Telegraph*, August 30, 2022, https://www.telegraph.co.uk/world-news/2022/08/30/ukraine-deploys-dummy-himars-trick-russian-forces/.
139. "Inflatable Tanks: Inside the Company That Makes Decoy Armaments," Manufacturing.net, March 6, 2023, https://www.manufacturing.net/operations/news/22751105/inflatable-tanks-inside-the-company-that-makes-decoy-armaments.
140. Katyanna Quach, "You Only Need Pen and Paper to Fool This OpenAI Computer Vision Code. Just Write Down What You Want It to See," *The Register*, March 5, 2021, https://www.theregister.com/2021/03/05/openai_writing_attack/.
141. Sean J. A. Edwards, *Swarming on the Battlefield: Past, Present, and Future* (Santa Monica, CA: RAND Corporation, 2000); John Arquilla and David Ronfeldt, *Swarming & the Future of Conflict* (Santa Monica, CA: RAND Corporation, 2000).
142. Arquilla and Ronfeldt, *Swarming & the Future of Conflict*, 8–9.

CHAPTER 4

143. Cade Metz, "One Genius' Lonely Crusade to Teach a Computer Common Sense," *Wired*, March 24, 2016, https://www.wired.com/2016/03/doug-lenat-artificial-intelligence-common-sense-engine/.
144. Micah Zenko, "Millennium Challenge: The Real Story of a Corrupted Military Exercise and Its Legacy," *War on the Rocks*, November 5, 2015, https://warontherocks.com/2015/11/millennium-challenge-the-real-story-of-a-corrupted-military-exercise-and-its-legacy/; Francis Horton, "The Lost Lesson of Millennium Challenge 2002, the Pentagon's Embarrassing Post-9/11 War Game," *Task and Purpose*, November 6, 2019, https://taskandpurpose.com/news/millenium-challenge-2002-stacked-deck/.

145. Robert C. Harney, "Broadening the Trade Space in Designing for Warship Survivability," *Naval Engineers Journal* 122, no. 1, November 2010, 49–63; William N. Reynolds and Peter A. Withers, "Morphological Analysis for Rapid, Low-Cost, Collaborative, Strategic Trade Space Analysis," Albuquerque, New Mexico: Least Squares Software, August 2011.
146. Stew Magnusen, "DARPA Pushes 'Mosaic Warfare' Concept," *National Defense* 103, November 1, 2018, 18–19.
147. K. J. Rawson and E. C. Tupper, *Basic Ship Theory*, 5th ed. (Oxford, UK: Butterworth Heinemann, 2001), 623.
148. Rawson and Tupper, *Basic Ship Theory*, 645.
149. Richard M. Ogorkiewicz, *Technology of Tanks* (Surrey, UK: Jane's Information Group, 1991), 384.
150. Elad Moisseiev and Gad Dotan, "Negative g-Force Ocular Trauma Caused by a Rapidly Spinning Carousel," *Case Reports in Opthamology* 4, No. 3, September–December 2013, 180–183.
151. Mark F. Cancian, Matthew Cancian, and Eric Heginbotham, *The First Battle of the Next War: Wargaming a Chinese Invasion of Taiwan* (Washington, DC: Center for Strategic and International Studies [CSIS], January 2023), 5.
152. Patrick Tucker and Jacqueline Feldscher, "As China, Taiwan Tensions Flare, US Faces Shrinking Window to Deter Conflict," *Defense One*, August 8, 2022, https://www.defenseone.com/threats/2022/08/china-taiwan-tensions-flare-us-faces-shrinking-window-deter-conflict/375514/.
153. Sebastien Roblin, "Black Sea Drone War: How a Country with No Warships Has Russia's Navy on the Run," *Inside Unmanned Systems*, October 19, 2023, https://insideunmannedsystems.com/black-sea-drone-war-how-a-country-with-no-warships-has-russias-navy-on-the-run/.
154. Dave Majumdar, "The US Navy's Great Littoral Combat Ship Reboot is Here," *The National Interest*, September 9, 2016, https://nationalinterest.org/blog/the-buzz/the-us-navys-great-littoral-combat-ship-reboot-here-17656; David Axe, "The Littoral Combat Ship Can't Fight—And the U.S. Navy Is Finally Coming to Terms With It," *Forbes*, May 20, 2021, https://www.forbes.com/sites/davidaxe/2021/05/20/the-littoral-combat-ship-cant-fight-the-us-navy-is-finally-coming-to-terms-with-it/?sh=273f72ba2587.
155. Michael Stott, "Deadly New Russian Weapon Hides in Shipping Container," Reuters, April 26, 2010, https://www.reuters.com/article/us-russia-weapon-idustre63p2xb20100426; Mark Vermylen, "Israel's Long-Range Artillery Weapon System (LORA)," Missile Defense Advocacy Alliance, June

2017, https://missiledefenseadvocacy.org/israels-long-range-artillery-weapon-system-lora/; Raul (Pete) Pedrozo, "China's Container Missile Deployments Could Violate the Law of Naval Warfare," *International Law Studies* 97, 2021, 1160–1170, https://digital-commons.usnwc.edu/cgi/viewcontent.cgi?article=2982&context=ils; Oliver Parken, "Iran Fires Ballistic Missile from a Shipping Container at Sea," *The War Zone*, February 14, 2024, https://www.twz.com/news-features/iran-fires-ballistic-missile-from-a-shipping-container-at-sea.

156. Vermylen, "Israel's Long-Range Artillery Weapon System (LORA)."
157. Joseph Trevithick, "Navy Unveils Truck-Mounted SM-6 Missile Launcher in European Test," *The War Zone*, September 14, 2022, https://www.thedrive.com/the-war-zone/navy-unveils-truck-mounted-sm-6-missile-launcher-in-european-test; Zach Abdi, "US Navy and Army's MK 70 Payload Delivery Systems Stretch Their Wings," *Naval News*, September 25, 2023, https://www.navalnews.com/naval-news/2023/09/u-s-navy-and-army-mk-70-pds-stretch-their-wings/.
158. Joseph Trevithick, "Shipping Container Launcher Packing 126 Kamikaze Drones Hits the Market," *The War Zone*, June 17, 2024, https://www.twz.com/land/shipping-container-launcher-packing-126-kamikaze-drones-hits-the-market.
159. J. P. Lawrence, "Navy's 'Influx' of Aquatic and Aerial Drones Tested in the Middle East," *Stars and Stripes*, December 1, 2022, https://www.stripes.com/branches/navy/2022-12-01/navy-drone-boats-bahrain-8260601.html.
160. "Bas 90—Air Base System 90, Swedish Air Force (1986) [English Subtitles Available]," June 4, 2017, https://www.youtube.com/watch?v=MNak9lB_q00.
161. Andrew Layton, "Historic Highway Landing Advances Agile Combat Employment," Air Force Reserve Command, July 4, 2022, https://www.afrc.af.mil/News/Article-Display/Article/3083909/historic-highway-landing-advances-agile-combat-employment/.
162. Lieutenant General Tony D. Bauernfeind, "Ready to Compete, Fight, and Win in the Indo-Pacific." (Air Force Association Air, Space, and Cyber Conference, National Harbor, Maryland, September 13, 2023.)
163. Sydney J. Freedberg, "'Unmanned' drones take too many humans to operate, says top Army aviator," Breaking Defense, February 27, 2023, https://breakingdefense.com/2023/02/unmanned-drones-take-too-many-humans-to-operate-says-top-army-aviator/.

164. *The Role of Autonomy in DoD Systems* (Washington, DC: Defense Science Board, Office of the Under Secretary of Defense for Acquisition, Technology, and Logistics, July 2012), 57.
165. Nick Paton Walsh, Florence Davey-Attlee, Kostya Gak, and Brice Laine, "Ukrainian Team Uses Thermal Cameras in Hunt for Russian Threat," CNN, August 15, 2023, https://edition.cnn.com/europe/live-news/russia-ukraine-war-news-08-15-23/h_174973601d0100674d5e8066a20a6ae8.
166. Ian Johnston and Rob McAuley, *The Battleships* (St. Paul, Minnesota: MBI Publishing Company, 2000), 180.

CHAPTER 5

167. David Barno and Nora Bensahel, "The Future of the Army: Today, Tomorrow, and the Day after Tomorrow" (Washington, DC: Atlantic Council, September 2016), 24, 28, https://www.atlanticcouncil.org/in-depth-research-reports/report/the-future-of-the-army-2/.
168. George M. Dougherty, "Ground Combat Overmatch Through Control of the Atmospheric Littoral," *Joint Force Quarterly* 94, 3rd Quarter 2019, 64–73; George M. Dougherty, "Control of the Atmospheric Littoral: A Future Doctrinal Framework for Unmanned Systems in Ground Combat," presented at the Naval Counter-Improvised Threat Knowledge Network Symposium, virtual, January 7, 2021, https://apps.dtic.mil/sti/pdfs/AD1120302.pdf; George M. Dougherty, "Fostering Emerging Robotics Autonomy Doctrine to Guide Future Force Design," presented at the U.S. Army Robotics Portfolio Integration Meeting (ARPIM), virtual, March 24, 2021.
169. Maximilian K. Bremer and Kelly A. Grieco, "The Air Littoral: Another Look," *Parameters* 51, no. 4 (Winter 2021), 67–80, doi:10.55540/0031-1723.3092; Jim E. Rainey and James K. Greer, "Land Warfare and the Air-Ground Littoral," *Army Aviation Magazine*, December 31, 2023, 14–17.
170. Wayne P. Hughes, "Build a Green-Water Fleet," U.S. Naval Institute Proceedings 144, no. 6 (June 2018), https://www.usni.org/magazines/proceedings/2018/june/build-green-water-fleet.
171. Ben Watson, "The Drones of ISIS," *Defense One*, January 12, 2017, https://www.defenseone.com/technology/2017/01/drones-isis/134542/.
172. Pablo Chovil, "Air Superiority Under 2000 Feet: Lessons from Waging Drone Warfare Against ISIL," *War on the Rocks*, May 11, 2018, https://

warontherocks.com/2018/05/air-superiority-under-2000-feet-lessons-from-waging-drone-warfare-against-isil/.

173. Jules Hurst, "The Developing Fight for Tactical Air Control," *War on the Rocks*, March 28, 2019, https://warontherocks.com/2019/03/the-developing-fight-for-tactical-air-control/.

174. George M. Dougherty, "Control of the Atmospheric Littoral: A Potential Doctrine to Enable Overmatch in Maneuver Warfare," presentation to the combined Army Centers of Excellence, Ft. Benning, GA, July 28, 2021.

175. Dougherty, "Ground Combat Overmatch Through Control of the Atmospheric Littoral."

176. Jen Hudson, "US Army Taps Industry for Autonomous Drones to Resupply Troops," *Defense News*, January 15, 2021, https://www.defensenews.com/land/2021/01/15/us-army-taps-industry-for-autonomous-drones-to-resupply-troops/.

177. Alex Kushleyev, Daniel Mellinger, and Vijay Kumar, "Towards a Swarm of Agile Micro Quadrotors," *Autonomous Robots* 35, no. 4 (November 2013), 287–300, https://doi.org/10.1007/s10514-013-9349-9.

178. Evan Ackerman, "This Autonomous Quadrotor Swarm Doesn't Need GPS," *IEEE Spectrum*, December 27, 2017, https://spectrum.ieee.org/this-autonomous-quadrotor-swarm-doesnt-need-gps; Caltech, "Close-Proximity Flight of Sixteen Quadrotor Drones," July 14, 2020, https://www.youtube.com/watch?v=geJt8PFZ-Fk.

179. Dougherty, "Control of the Atmospheric Littoral: A Potential Doctrine to Enable Overmatch in Maneuver Warfare"; George M. Dougherty, "Control of the Atmospheric Littoral: An Emerging Conceptual Framework for Robotics in Maneuver Warfare," in Maneuver Warfighter Conference, Ft. Benning, GA, February 15-17, 2022, https://www.youtube.com/watch?v=PbOMHFEE3OU.

180. "Army Readies Charging Port for Autonomous Drone Swarms," October 7, 2020, https://www.army.mil/article/239742/army_readies_charging_port_for_autonomous_drone_swarms.

181. Evan Ackerman, "Every Quadrotor Needs This Amazing Failsafe Software," *IEEE Spectrum*, March 4, 2014, https://spectrum.ieee.org/automaton/robotics/drones/every-quadrotor-needs-this-amazing-failsafe-software; Terry Jarrell, "Motor Failure Does Not Have to Crash Your Quadcopter," Aircraft Owners and Pilots Association, April 26, 2019, https://www.aopa

.org/news-and-media/all-news/2019/april/26/this-failsafe-can-save-the-day-if-your-drone-loses-a-motor.

182. Javier Chagoya, "NPS, Academic Partners Take to the Skies in First-Ever UAV Swarm Dogfight," Naval Postgraduate School, February 22, 2017, https://nps.edu/-/nps-academic-partners-take-to-the-skies-in-first-ever-uav-swarm-dogfight.

CHAPTER 6

183. David Hambling, "AI Thrashes Human Fighter Pilot 5-0 in Simulated F-16 Dogfights," *New Scientist*, August 25, 2020, https://www.newscientist.com/article/2252760-ai-thrashes-human-fighter-pilot-5-0-in-simulated-f-16-dogfights/.

184. "AlphaDogfight Trials Foreshadow Future of Human-Machine Symbiosis," Defense Advanced Research Projects Agency, August 26, 2020, https://www.darpa.mil/news-events/2020-08-26.

185. Stephen Losey, "U.S. Air Force Plans Self-Flying F-16s to Test Drone Wingman Tech," *Defense News*, March 28, 2023, https://www.defensenews.com/air/2023/03/28/us-air-force-plans-self-flying-f-16s-to-test-drone-wingmen-tech/.

186. Khari Johnson, "Drone Racing League Launches $2 Million Autonomous Drone Competition," *VentureBeat*, September 5, 2019, https://venturebeat.com/ai/drone-racing-league-launches-2-million-autonomous-drone-competition/.

187. Stephen Babcock, "Autonomous Drone Racing Puts AI Behind the Controllers," *Technical.ly*, November 18, 2019, https://technical.ly/software-development/autonomous-drone-racing-ai-artificial-intelligence-lockheed-martin-league/.

188. Christophe De Wagter, Federico Paradea-Vallés, Nilay Sheth, and Guido C. H. E. de Croon, "The Artificial Intelligence Behind the Winning Entry to the 2019 AI Robotic Racing Competition," *Field Robotics* 2, September 30, 2021, 1263–1290, https://arxiv.org/pdf/2109.14985.pdf.

189. Christophe De Wagter, Federico Paradea-Vallés, Nilay Sheth, and Guido C. H. E. de Croon, "Learning Fast in Autonomous Drone Racing," *Nature Machine Intelligence* 3, no. 10, October 2021, 923.

190. Benj Edwards, "High-Speed AI Drone Beats World-Champion Racers for the First Time," *Ars Technica*, August 31, 2023, https://arstechnica

.com/information-technology/2023/08/high-speed-ai-drone-beats-world-champion-racers-for-the-first-time/.

191. Interview of Former USS Missouri Executive Officer, Lead Sheet #14246, Washington, DC: Office of the Special Assistant for Gulf War Illnesses, US Department of Defense, January 23, 1998, https://gulflink.health.mil/du_ii/du_ii_refs/n52en228/8023_034_0000001.htm; "Missile Attack on the Battleship USS Missouri," *wwiiafterwwii: WWII Equipment Used After the War* (blog), July 21, 2019, https://wwiiafterwwii.wordpress.com/2019/07/21/missile-attack-on-battleship-uss-missouri/.
192. John K. Hawley, "Patriot Wars: Automation and the Patriot Air and Missile Defense System," Washington, DC: Center for a New American Security, January 25, 2017, 6–8. https://www.cnas.org/publications/reports/patriot-wars.
193. Sam Lagrone, "U.S. Super Hornet Shot Down Over Red Sea in Friendly Fire Incident; Aviators Safe," *U.S. Naval Institute News*, December 21, 2024, https://news.usni.org/2024/12/21/u-s-super-hornet-shot-down-over-red-sea-in-friendly-fire-incident-aviators-safe.
194. Paul Scharre, *Four Battlegrounds: Power in the Age of Artificial Intelligence* (New York: W. W. Norton and Company, 2023), 231.
195. Katyanna Quach, "You Only Need Pen and Paper to Fool This OpenAI Computer Vision Code. Just Write Down What You Want It to See," *The Register*, March 5, 2021, https://www.theregister.com/2021/03/05/openai_writing_attack/.
196. Alex Krizhevsky, Ilya Sutskever, and Geoffrey E. Hinton, "ImageNet Classification with Deep Convolutional Neural Networks," Communications of the Association for Computing Machinery 60, December 3, 2012, 84–90, DOI:10.1145/3065386; Dave Gershgorn, "The Inside Story of How AI Got Good Enough to Dominate Silicon Valley," *Quartz*, June 18, 2018, https://qz.com/1307091/the-inside-story-of-how-ai-got-good-enough-to-dominate-silicon-valley.
197. David Gelles, "A.I.'s Insatiable Appetite for Energy," *New York Times*, July 11, 2024, https://www.nytimes.com/2024/07/11/climate/artificial-intelligence-energy-usage.html.
198. Alex Hughes, "ChatGPT: Everything You Need to Know About OpenAI's GPT-4 Tool," *BBC Science Focus*, September 25, 2023, https://www.sciencefocus.com/future-technology/gpt-3.
199. Carl von Clausewitz, *On War* (Princeton, New Jersey: Princeton University Press, 2008), 101–102, 119–120.

200. Cade Metz, "One Genius' Lonely Crusade to Teach a Computer Common Sense," *Wired*, March 24, 2016, https://www.wired.com/2016/03/doug-lenat-artificial-intelligence-common-sense-engine/.

201. Doug Lenat, "Sometimes the Veneer of Intelligence Is Not Enough," *CognitiveWorld*, July 2, 2021, https://cognitiveworld.com/articles/sometimes-veneer-intelligence-not-enough.

202. Augustin M. Prentiss, *Chemicals in War: A Treatise on Chemical Warfare* (New York and London: McGraw-Hill, 1937), 78.

203. Dan Coles, "The WW2 Anti-Tank Bomb Dogs Which Blew Up Their Own Side's Tanks Instead of the Enemy's," *Express* (UK), August 6, 2023, https://www.express.co.uk/news/uk/1798899/Soviet-dogs-anti-Nazi-tank.

204. Kathleen L. Mosier and Linda J. Skitka, "Human Decision Makers and Automated Decision Aids: Made for Each Other?" in R. Parasuraman, and M. Mouloua, eds., *Automation and Human Performance: Theory and Applications* (Hillsdale, NJ: Lawrence Erlbaum Associates, Inc, 1996), 201–220.

205. Kate Goddard, Abdul Roudsari, and Jeremy C. Wyatt, "Automation Bias: A Systematic Review of Frequency, Effect Mediators, and Mitigators," *Journal of the American Medical Informatics Association* 19, no. 1, January–February 2012, 121–127.

206. John Christianson, DI Cooke, and Courtney Stiles Herdt, "Miscalibration of Trust in Human Machine Teaming, *War on the Rocks*, March 8, 2023, https://warontherocks.com/2023/03/miscalibration-of-trust-in-human-machine-teaming/.

207. Nataly Delcid, "Is Google's AI Sentient? Stanford AI Experts Say That's 'Pure Clickbait.'" *Stanford Daily*, August 2, 2022, https://stanforddaily.com/2022/08/02/is-googles-ai-sentient-stanford-ai-experts-say-thats-pure-clickbait/.

208. Elliot Leavy, "Full Transcript: Google Engineer Talks to 'Sentient' Artificial Intelligence," *AI Data & Analytics Network*, June 14, 2022, https://www.aidataanalytics.network/data-science-ai/news-trends/full-transcript-google-engineer-talks-to-sentient-artificial-intelligence-2.

209. Brad Darrach, "Meet Shaky, the First Electronic Person: The Fascinating and Fearsome Reality of a Machine with a Mind of Its Own," *Life*, November 20, 1970, 58B–68.

210. James Whale, dir. *Frankenstein*, Los Angeles, CA: Universal Pictures, 1931, https://www.youtube.com/watch?v=1qNeGSJaQ9Q.

211. James J. Hall, *American Kamikaze* (Titusville, FL: J. Bryant, Ltd., 1984), 161.

212. Karlheinz Münch, *Combat History of Schwere Panzerjäger Abteilung 653*, translated by Bo H. Friesen (Winnipeg, Canada: J J Fedorowicz Publishers, 1997), 53.
213. Jeremy Hsu, "Robot Funerals Reflect Our Humanity," *Discover*, March 15, 2015, https://www.discovermagazine.com/technology/robot-funerals-reflect-our-humanity.
214. Ray Kurzweil, *The Age of Spiritual Machines* (New York: Penguin Books, 1999), 66.
215. Shanee Honig and Tal Oron-Gilad, "Understanding and Resolving Failures in Human-Robot Interaction: Literature Review and Model Development," *Frontiers in Psychology* 9, Article 861, June 2018, doi:10.3389/fpsyg.2018.00861.
216. Heather R. Penney, Maj. Christopher Olsen, and Lt. Gen David A. Deptula (ret), *Beyond Pixie Dust: A Framework for Understanding and Developing Autonomy in Unmanned Aircraft* (Arlington, Virginia: Mitchell Institute for Aerospace Studies), February 2022.
217. Ann-Renee Blais and Megan Thompson, *The Trust in Teams and Trust in Leaders Scale: A Review of Their Psychometric Properties and Item Selection*, TM 2009-161 (Toronto: Defense Research and Development Canada), September 2009.

CHAPTER 7

218. Dima Adamsky, *The Culture of Military Innovation: The Impact of Cultural Factors on the Revolution in Military Affairs in Russia, the US, and Israel* (Stanford, California: Stanford University Press, 2010), 78–81, 91.
219. Gen. Charles C. Krulak, "The Strategic Corporal: Leadership in the Three Block War," *Leatherneck: The Marine Corps Gazette*, January 1999, 14.
220. *The Joint Team* (Washington, DC: Department of the Air Force, 2022), 9.
221. *Joint Publication 5-0: Joint Operations Planning* (Washington, DC: Joint Chiefs of Staff, 2011), III-39, https://jfsc.ndu.edu/Portals/72/Documents/JC2IOS/DOPC/JP%205_0%20Joint%20Planning.pdf.
222. "Joint Light Tactical Vehicle (JLTV) Analysis of Alternatives (AoA)," in *49th Army Operations Research Symposium*, Fort Lee, Virginia, October 13–14, 2010, https://www.slideserve.com/zev/joint-light-tactical-vehicle-jltv-analysis-of-alternatives-aoa.

223. "US Navy Tests Raytheon SM-6 Missile's Surface-to-Surface Engagement Capability," *Naval Technology*, March 7, 2016, https://www.naval-technology.com/news/newsus-navy-tests-raytheon-sm-6-missiles-surface-to-surface-engagement-capability-4832692/.
224. Sebastien Roblin, "The Air Force Is Making a Big Bet on Stormbreaker Bombs," *Popular Mechanics*, April 8, 2023, https://www.popularmechanics.com/military/aviation/a43483654/air-force-buying-stormbreaker-smart-bombs/; Thomas Newdick, "F-35 to Get Meteor, SPEAR 3 Missiles by 'End of Decade,'" *The WarZone*, January 22, 2024, https://www.twz.com/f-35-to-get-meteor-spear-3-missiles-by-end-of-decade.
225. Paolo Valpolini, "MBDA Germany Enforcer Lightweight Missile Is Ready for Production and to Generate New Variants," *European Defence Review*, October 31, 2023, https://www.edrmagazine.eu/mbda-germany-enforcer-lightweight-missile-is-ready-for-production-and-to-generate-new-variants.
226. Robert Gates, speech to the US Military Academy, West Point, NY, February 25, 2011.
227. General William C. Westmoreland, Congressional testimony, October 16, 1969, reprinted in *Air War: The Third Indochina War* (Washington, DC: Indochina Resource Center, March 1972), 13.
228. Thomas X. Hammes, *The Sling and the Stone: On War in the 21st Century* (Minneapolis, Minnesota: Zenith Press, 2006), 208–212.
229. Harry G. Summers, Jr., *On Strategy: The Vietnam War in Context* (Carlisle Barracks, Pennsylvania: US Army War College, 1982), 1.
230. Steven Pressfield, *The Warrior Ethos* (Reseda, California: Black Irish Books, 2011), 14.
231. Hammes, *The Sling and the Stone*, 14–15.
232. Sean McFate, *The New Rules of War: How America Can Win—Against Russia, China, and Other Threats* (New York: William Morrow, 2019), 64–65.
233. Tzu-Chieh Hung and Tzu-Wei Hung, "How China's Cognitive Warfare Works: A Frontline Perspective of Taiwan's Anti-Disinformation Wars," *Journal of Global Security Studies* 7, no. 4, December 2022, https://doi.org/10.1093/jogss/ogac016.
234. Valery Gerasimov, "The Development of Military Strategy Under Contemporary Conditions. Tasks for Military Science," March 2019, translated by Harold Orenstein and Timothy Thomas, *Military Review*, November 2019,

https://www.armyupress.army.mil/Portals/7/Army-Press-Online-Journal/documents/2019/Orenstein-Thomas.pdf.

235. Joseph Trevithick, "Massive Drone Swarm Over Strait Decisive in Taiwan Conflict Wargames," *The WarZone*, May 19, 2022, https://www.twz.com/massive-drone-swarm-over-strait-decisive-in-taiwan-conflict-wargames.
236. Mark A. Gunzinger, Lawrence A. Stutzreim, and Bill Sweetman, *The Need for Collaborative Combat Aircraft for Disruptive Air Warfare* (Arlington, Virginia: Mitchell Institute for Aerospace Studies, February 2024), 36–37.
237. William Claiborne, "Arabs Agree on Force to Defend Saudis," *Washington Post*, August 11, 1990.
238. Alia Shoaib, "Inside the Elite Drone Unit Founded by Volunteer IT Experts: 'We Are All Soldiers Now,'" *Business Insider*, April 9, 2022, https://www.businessinsider.com/inside-the-elite-ukrainian-drone-unit-volunteer-it-experts-2022-4.
239. Julian Borger, "The Drone Operators Who Halted Russian Convoy Headed for Kyiv: Special IT Force of 30 Soldiers on Quad Bikes Is Vital Part of Ukraine's Defence, But Forced to Crowdfund for Supplies," *The Guardian*, March 28, 2022, https://www.theguardian.com/world/2022/mar/28/the-drone-operators-who-halted-the-russian-armoured-vehicles-heading-for-kyiv.
240. Luke Harding and Peter Sauer, "Ukraine Levels Up the Fight with Drone Strikes Deep into Russia," *The Guardian*, January 27, 2024, https://www.theguardian.com/world/2024/jan/27/ukraine-levels-up-the-fight-with-drone-strikes-deep-into-russia.
241. "Ukrainian Saboteurs Behind Attacks Inside Russia, Reports Say," *Voice of America News*, August 22, 2023, https://www.voanews.com/a/russia-downs-ukrainian-drones-in-moscow-region-/7234981.html; Peter Suciu, "Ukrainian Video Shows Destruction of Il-76 Aircraft—Contradicting the Kremlin's Claims," *Forbes*, September 1, 2023, https://www.forbes.com/sites/petersuciu/2023/09/01/ukrainian-video-shows-destruction-of-il-76-aircraft--contradicting-the-kremlins-claims/?sh=3e4f29318f4b.
242. Seth J. Frantzman, "Are Air Defense Systems Ready to Confront Drone Swarms?" *Defense News*, September 26, 2019, https://www.defensenews.com/global/mideast-africa/2019/09/26/are-air-defense-systems-ready-to-confront-drone-swarms/.

243. Geoff Brumfiel, "What We Know About the Attack on Saudi Oil Facilities," NPR, September 19, 2019, https://www.npr.org/2019/09/19/762065119/what-we-know-about-the-attack-on-saudi-oil-facilities.
244. Asif Shahzad and Gibran Naiyyar Peshimam, "Pakistan Strikes Inside Iran Against Militant Targets, Stokes Regional Tension," Reuters, January 18, 2024, https://www.msn.com/en-gb/news/world/pakistan-strikes-inside-iran-against-militant-targets-stokes-regional-tension/ar-AA1naXxG.
245. Joseph L. Votel and Eero R. Keravuori, "The By-With-Through Operational Approach," *Joint Force Quarterly* 89, 2nd Quarter, 2018, 40–47, https://ndupress.ndu.edu/Media/News/News-Article-View/Article/1491891/the-by-with-through-operational-approach/.
246. Bruce Riedel, "The Mess in Afghanistan," Brookings, March 4, 2020, https://www.brookings.edu/articles/the-mess-in-afghanistan/.

CHAPTER 8

247. Niccolò Machiavelli, *The Art of War*, translated by Christopher Lynch (Chicago: University of Chicago Press, 2005), 74.
248. Francesco Guicciardini, *Opere Inedite di Francesco Guicciardini III: Storia Fiorentina* (Firenze: Barbera, Bianchi and Company, 1859), 105.
249. Max Boot, *War Made New: Technology, Warfare, and the Course of History* (New York: Gotham, 2006), 23–24.
250. Noel Perrin, *Giving Up the Gun: Japan's Reversion to the Sword, 1543–1879* (Boston: David R. Godine, 1979); Alexander Astroth, "The Decline of Japanese Firearm Manufacturing and Proliferation in the Seventeenth Century," *Emory Endeavors in History* 5, 2013, 136–148, http://history.emory.edu/home/documents/endeavors/volume5/gunpowder-age-v-astroth.pdf.
251. Clayton M. Christensen, *The Innovator's Dilemma: When New Technologies Cause Great Firms to Fail* (Cambridge, Massachusetts: Harvard Business School Press, 1997).
252. "Ukraine to Produce a Million FPV Drones Next Year—Minister," Reuters, December 20, 2023, https://www.reuters.com/world/europe/ukraine-produce-million-fpv-drones-next-year-minister-2023-12-20/; Inder Singh Bisht, "Ukraine Producing More Drones Than the State Can Buy," *The Defense Post*, January 11, 2024, https://www.thedefensepost.com/2024/01/11/ukraine-producing-drones-state/.

253. Pesha Magid, "Turkish War Drone Factory to Open in Ukraine with Plans to Supply to 30 Countries," *The Independent*, February 6, 2024, https://www.independent.co.uk/news/world/europe/ukraine-drones-baykar-kyiv-russia-b2491543.html.
254. John Krzyzaniak, "The Private Companies Propelling Iran's Drone Industry," Iran Watch, November 29, 2023, https://www.iranwatch.org/our-publications/articles-reports/private-companies-propelling-irans-drone-industry.
255. Declan Walsh, "Foreign Drones Tip the Balance in Ethiopia's Civil War," *New York Times*, December 20, 2021, https://www.nytimes.com/2021/12/20/world/africa/drones-ethiopia-war-turkey-emirates.html.
256. Zecharias Zelalem, "'Collective Punishment': Ethiopia Drone Strikes Target Civilians in Amhara," *Al Jazeera*, December 29, 2023, https://www.aljazeera.com/features/2023/12/29/collective-punishment-ethiopia-drone-strikes-target-civilians-in-amhara.
257. Wilson McMakin, "Moroccan Drone Strikes Force Sahrawi from Their Homes," *The New Humanitarian*, May 17, 2023, https://www.thenewhumanitarian.org/news-feature/2023/05/17/morocco-sahrawi-drone-attacks.
258. Patrick Wintour and Ghaith Abdul-Ahad, "Iraq Labels U.S. 'Factor of Instability' After Three Killed in Drone Attack," *The Guardian*, February 8, 2024, https://www.theguardian.com/world/2024/feb/07/baghdad-drone-strike-iraq-iran-militia.
259. "International Players Behind Libya's Drone War," *Times Aerospace*, December 5, 2019, https://www.timesaerospace.aero/features/international-players-behind-libyas-drone-war; Léo Péria-Peigné, *TB2 Bayraktar: Big Strategy for a Little Drone*, Paris: French Institute of International Relations Security Studies Center, April 17, 2023, 4, https://www.ifri.org/sites/default/files/atoms/files/peria-peigne_tb2_bayraktar_2023.pdf.
260. Luis Chaparro, "The Wrong People Just Got Their Hands on an Elite Drone Unit," *The Daily Beast*, August 22, 2023, https://www.thedailybeast.com/mexicos-jalisco-new-generation-cartel-just-created-an-elite-drone-unit.
261. "Drug Cartels Are Sharply Increasing Use of Bomb-Dropping Drones, Mexican Army Says," CBS News, August 23, 2023, https://www.cbsnews.com/news/drug-cartels-more-bomb-dropping-drones-mexico-army/.
262. Christopher D. Booth and Walker D. Mills, "Unfurl the Banner! Privateers and Commerce Raiding of China's Merchant Fleet in Developing Markets,"

War on the Rocks, February 18, 2021, https://warontherocks.com/2021/02/unfurl-the-banner-privateers-and-commerce-raiding-of-chinas-merchant-fleet-in-developing-markets/.

263. "Death Rate in Wars, World—Correlates of War," Our World in Data, 2020, https://ourworldindata.org/grapher/death-rate-in-wars-correlates-of-war; Zack Beauchamp, "600 Years of War and Peace, in One Amazing Chart," *Vox*, June 24, 2015, https://www.vox.com/2015/6/23/8832311/war-casualties-600-years.
264. Mark Jacobsen, "The Strategic Implications of Non-State #Warbots," The Strategy Bridge, December 15, 2017, https://thestrategybridge.org/the-bridge/2017/12/15/the-strategic-implications-of-non-state-warbots.
265. Sean McFate, *The New Rules of War: Victory in an Age of Durable Disorder* (New York: William Morrow, 2019), 8–9.
266. McFate, *The New Rules of War*, 198–203.
267. Max Roser, "Global Deaths in Conflicts Since the Year 1400," Our World in Data, 2015.

CHAPTER 9

268. "Ukraine: Deadly Mariupol Theatre Strike 'A Clear War Crime' by Russian Forces—New Investigation," Amnesty International, June 30, 2022, https://www.amnesty.org/en/latest/news/2022/06/ukraine-deadly-mariupol-theatre-strike-a-clear-war-crime-by-russian-forces-new-investigation/.
269. Lori Hinnant, Mstyslav Chernov, and Vasilisa Stepanenko, "AP Evidence Points to 600 Dead in Mariupol Theater Airstrike," Associated Press, May 4, 2022, https://apnews.com/article/russia-ukraine-entertainment-europe-donetsk-59fbc059a9fe9b5825841ff25025b5cc.
270. Hugo Bachega and Orysia Khimiak, "A Bomb Hit This Theatre Hiding Hundreds—Here's How One Woman Survived," BBC News, March 22, 2022, https://www.bbc.com/news/world-europe-60835106.
271. Bachega and Khimiak, "A Bomb Hit This Theatre Hiding Hundreds"; "Ukraine: Deadly Mariupol Theater Strike," Amnesty International.
272. Jay Winter and Blaine Baggett, *The Great War and the Shaping of the 20th Century* (New York: Penguin Studio, 1996), 65–66.
273. Luke Harding, "'Horrendous' Rocket Attack Kills Civilians in Kharkiv as Moscow 'Adapts Its Tactics,'" *The Guardian*, February 28, 2022, https://

www.theguardian.com/world/2022/feb/28/ukraine-several-killed-by-russian-rocket-strikes-in-civilian-areas-of-kharkiv.

274. Marcus Walker and Yaroslav Trofimov, "Russia Tried to Freeze Ukraine. Here's How It Survived the Winter," *Wall Street Journal*, March 20, 2023, https://www.wsj.com/articles/how-ukraine-survived-russias-mission-to-turn-off-the-lights-winter-is-over-and-were-still-here-8668c5f5.
275. "Russian Offensive Campaign Assessment, November 15," Institute for the Study of War, November 15, 2022, https://www.understandingwar.org/backgrounder/russian-offensive-campaign-assessment-november-15.
276. "Syria/Russia: Strategy Targeted Civilian Infrastructure," Human Rights Watch, October 15, 2020, https://www.hrw.org/news/2020/10/15/syria/russia-strategy-targeted-civilian-infrastructure; Janine Di Giovanni, "Vladimir Putin's Inhumane Blueprint to Terrorize Civilians in Chechnya, Syria—and Now Ukraine," *Vanity Fair*, February 23, 2023, https://www.vanityfair.com/news/2023/02/vladimir-putin-chechnya-syria-ukraine.
277. Zecharias Zelalem, "'Collective Punishment': Ethiopia Drone Strikes Target Civilians in Amhara," *Al Jazeera*, December 29, 2023, https://www.aljazeera.com/features/2023/12/29/collective-punishment-ethiopia-drone-strikes-target-civilians-in-amhara.
278. Bir Moghrein, "Moroccan Drone Strikes Force Sahrawi from Their Homes," *The New Humanitarian*, May 17, 2023, https://www.thenewhumanitarian.org/news-feature/2023/05/17/morocco-sahrawi-drone-attacks.
279. David Hambling, "Drones May Have Attacked Humans Fully Autonomously for the First Time," *New Scientist*, May 27, 2021, https://www.newscientist.com/article/2278852-drones-may-have-attacked-humans-fully-autonomously-for-the-first-time/.
280. Randall Jarrell, "Losses" in *Losses: Poems by Randall Jarrell* (New York: Harcourt, Brace and Company, 1948).
281. Eyal Press, "The Wounds of the Drone Warrior," *New York Times Magazine*, June 17, 2018, 30–37, 47, 49.
282. Paul Lushenko and Shyam Raman, "Biden Can Reduce Civilian Casualties During U.S. Drone Strikes. Here's How," Brookings, January 19, 2022, https://www.brookings.edu/articles/biden-can-reduce-civilian-casualties-during-us-drone-strikes-heres-how/.
283. Alex Horton and Serhii Korolchuk, "Ukraine's Main Attack Drone Gives an Intimate Look at Battle," *Washington Post*, October 15, 2023, A1; "Darwin's

War: Inside the Secret Bunker of Ukraine's Ace FPV Drone Pilot," Scripps News, May 19, 2024, https://www.youtube.com/watch?v=WipqeFgzdTc.

284. "Military and Killer Robots," Campaign to Stop Killer Robots, accessed March 2, 2024, https://www.stopkillerrobots.org/military-and-killer-robots/.

285. US Joint Chiefs of Staff, *Mission Command, Second Edition*, Suffolk, Virginia: Joint Staff J7, January 2020, https://www.jcs.mil/Portals/36/Documents/Doctrine/fp/missioncommand_fp_2nd_ed.pdf?ver=2020-01-13-083451-207; Air Force Doctrine Publication 1-1, *Mission Command*, August 14, 2023, https://www.doctrine.af.mil/Portals/61/documents/AFDP_1-1/AFDP%201-1%20Mission%20Command.pdf.

286. Air Force Doctrine Publication 1-1, *Mission Command*, 5.

287. Stanislaw Lem, *His Master's Voice*, translated by Michael Kandel (Evanston, Illinois: Northwestern University Press, 1999), 74–75.

288. Curtis E. LeMay with MacKinlay Kantor, *Mission with LeMay: My Story* (Garden City, NY: Doubleday, 1965), 564–565.

289. Cofer Black, Testimony to the US Senate Select Intelligence Committee, September 26, 2002, https://irp.fas.org/congress/2002_hr/092602black.html; "What Happens When the Gloves Come Off," Human Rights Watch, April 8, 2008, https://www.hrw.org/news/2008/04/08/what-happens-when-gloves-come.

290. David Petraeus and Michael O'Hanlon, "Take the Gloves Off Against the Taliban," *Wall Street Journal*, May 20, 2016, https://www.wsj.com/articles/take-the-gloves-off-against-the-taliban-1463783106.

291. Daniel R. Mahanty, "Take Off the Pentagon's Gloves in the ISIS War? Not So Fast," *Defense One*, February 15, 2017, https://www.defenseone.com/ideas/2017/02/take-pentagons-gloves-isis-war-not-so-fast/135454/.

292. Jane Goodall, *Through a Window: My Thirty Years with the Chimpanzees of Gombe* (New York: Houghton Mifflin, 1992), 77–79, 98–111; Simon Townsend, Katie Slocombe, Melissa Thompson, and Klaus Zuberbühler, "Female-Led Infanticide in Wild Chimpanzees," *Current Biology* 17, no. 10, May 2007, R355–R356; Michael Wilson, Christophe Boesch, Barbara Fruth, et al., "Lethal Aggression in *Pan* Is Better Explained by Adaptive Strategies than Human Impacts," *Nature* 513, September 2014, 414–417.

293. Anne Barnard, "Inside Syria's Torture Prisons," *New York Times*, May 12, 2019, A1.

CHAPTER 10

294. Tim Martin, "UK Drone Test Squadron Fails to Register a Single Test Since Forming in 2020: Procurement Minister," *Breaking Defense*, March 6, 2024, https://breakingdefense.com/2024/03/uk-drone-test-squadron-fails-to-register-a-single-test-since-forming-in-2020-procurement-minister/.
295. Gabriel Dominguez, "Recruitment Issues Undermining Japan's Military Buildup," *Japan Times*, January 2, 2023, https://www.japantimes.co.jp/news/2023/01/02/national/japan-sdf-recruitment-problems/.
296. Danielle Sheridan, "Navy Has So Few Sailors It Has to Decommission Ships," *The Telegraph*, January 4, 2024, https://www.telegraph.co.uk/news/2024/01/04/royal-navy-few-sailors-decommission-ships-new-frigates/.
297. Frank Kendall, secretary of the Air Force, "Department of the Air Force Operational Imperatives," March 3, 2022, https://www.af.mil/Portals/1/documents/2023SAF/OPERATIONAL_IMPARITIVES_INFOGRAPHIC.pdf.
298. John Gall, *Systemantics: How Systems Really Work and How They Fail* (New York: Quadrangle—New York Times Book Company, 1977), 71.
299. Clayton M. Christensen, *The Innovator's Dilemma: When New Technologies Cause Great Firms to Fail* (Cambridge, Massachusetts: Harvard Business School Press, 1997), 121, 138.
300. Walter J. Boyne, "How the Predator Grew Teeth," *Air Force* 92, no. 7, July 2009, 42–45.
301. Richard Whittle, *Predator: The Secret Origins of the Drone Revolution* (New York: Henry Holt and Company, 2014) 154–155; Headquarters USAFE, Special Order GD-09, January 16, 2001, https://nsarchive2.gwu.edu/NSAEBB/NSAEBB484/.
302. Alex Bierbauer and Mark Cooter with Michael Marks, *Never Mind, We'll Do It Ourselves: The Inside Story of How a Team of Renegades Broke Rules, Shattered Barriers, and Launched a Drone Warfare Revolution* (New York: Skyhorse Publishing, 2021), 85, 128.
303. Whittle, *Predator*, 245.
304. John A. Bonin, *Army Aviation Becomes an Essential Arm: From the Howze Board to the Modular Force, 1962–2004*, Ph.D. dissertation, Philadelphia, Pennsylvania: Temple University, 2006, 95; Virgil Ney, *Evolution of the U.S. Army Field Manual, Valley Forge to Vietnam*, Combat Operations Research Group Study (CORG-M-244) (Fort Belvoir, Virginia: Headquarters U.S. Army Combat Development Command, January 1966), 107.

305. David C. Aronstein and Albert C. Piccirillo, *Have Blue and the F-117A: Evolution of the "Stealth Fighter"* (Reston, Virginia: American Institute of Aeronautics and Astronautics, 1997), 153.
306. Zach Rosenberg, "US Air Force to Set Up Experimental Squadron to Refine Collaborative Combat Aircraft Force Structure," Janes, November 16, 2023, https://www.janes.com/osint-insights/defence-news/air/us-air-force-to-set-up-experimental-squadron-to-refine-collaborative-combat-aircraft-force-structure.
307. David Evans, "A Missile That Can Smash Tanks Barely Dents Army Brass," *Chicago Tribune*, May 13, 1988, 21; David C. Morrison, "FOG-M's Slow March from Basement to Battlefield," *Lasers & Optronics* 9 no. 1, January 1990, 22.
308. Paul Scharre, "The Perilous Coming Age of AI Warfare," *Foreign Affairs*, February 29, 2024, https://www.foreignaffairs.com/ukraine/perilous-coming-age-ai-warfare.
309. Oriana Pawlyk, "This Air Force Unit is Getting the Military's First Robot Dogs," *Military.com*, November 12, 2020, https://www.military.com/daily-news/2020/11/12/air-force-unit-getting-militarys-first-robot-dogs.html.
310. Michael Lemish, *Forever Forward: K-9 Operations in Vietnam* (Atglen, PA: Schiffer Military History, 2009), 37.
311. Lemish, *Forever Forward*, 30.
312. Cameron Ford, personal interview, January 28, 2020.
313. Mike Ritland with Gary Brozer, *Trident K-9 Warriors: My Tale from the Training Ground to the Battlefield with Elite Navy SEAL Canines* (New York: St. Martin's Press, 2013), 177–178.
314. Ritland and Brozer, *Trident K-9 Warriors*, 122.
315. Ritland and Brozer, *Trident K-9 Warriors*, 125–126.
316. Ritland and Brozer, *Trident K-9 Warriors*, 123–124.
317. Major Jordan T. Criss and Master Sergeant Andre G. Hernandez, 325th Security Forces Squadron, personal interview, April 23, 2021.
318. Criss and Hernandez, personal interview, April 23, 2021.
319. Cameron Ford, personal interview, January 28, 2020.
320. William A. Cohen, *Drucker on Leadership: New Lessons from the Father of Modern Management* (New York: John Wiley & Sons, 2009), 254.

INDEX

B

D

E

N

O

S

T

U

V

ABOUT THE AUTHOR

George M. Dougherty is a senior military leader, scientist, and business strategist. In his service as an active duty and reserve officer, he held the second-highest military position in US Air Force science and technology. He is coauthor of the Air Force's current science and technology strategy. He led technical programs and units and helped develop cutting-edge systems including the F-22 stealth fighter. In his civilian career, he has served as an executive in two international management consulting firms, helping the senior leaders of Fortune 500 corporations and high-growth technology and life sciences companies to navigate strategic change in their industries. Dougherty holds a BS and MS in engineering from the University of Virginia; a PhD in materials science and engineering from the University of California, Berkeley; and an MBA from Cornell University.